THE EXTREMITIES
Muscles and Motor Points

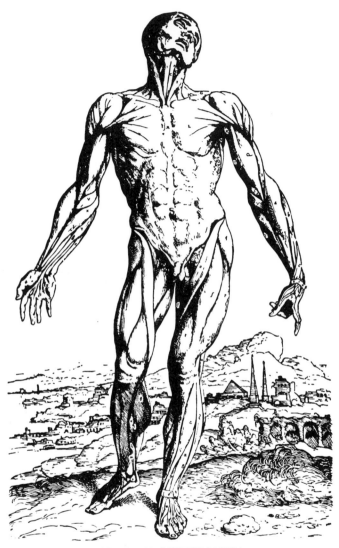

DE HUMANI CORPORIS FABRICA
ANDREAE VESALII
1568

THE EXTREMITIES
Muscles and Motor Points

JOHN H. WARFEL, Ph.D.
Associate Professor of Anatomy, State University of New York at Buffalo
School of Medicine, Buffalo, New York

Fifth Edition

110 Illustrations

LEA & FEBIGER

Philadelphia

Lea & Febiger
600 Washington Square
Philadelphia, PA 19106-4198
U.S.A.
(215) 922-1330

First Edition, 1945
Reprinted
June, 1946
April, 1948
September, 1950
January, 1952
October, 1953
November, 1954
September, 1956
February, 1958

Second Edition, 1960
Reprinted
August, 1963

Third Edition, 1967
Reprinted
March, 1970

Fourth Edition, 1974
Reprinted
April, 1976
May, 1978
January, 1981
November, 1982

Fifth Edition, 1985

Library of Congress Cataloging in Publication Data

Warfel, John H.
    The extremities: muscles and motor points.

    Includes index.
    1. Extremities (Anatomy)   2. Muscles—Innervation.
3. Myoneural junction.   I. Title.   [DNLM: 1. Extremities—
anatomy & histology—atlases.   2. Muscles—anatomy &
histology—atlases.   WE 17 W275e]
QM165.W27   1984   611'.738   84-12275
ISBN 0-8121-0965-1

Printed in the United States of America

Print Number: 6   5   4   3   2

PREFACE

The references in this fifth edition have been corrected to the 30th American Edition of *Gray's Anatomy* (Philadelphia, Lea & Febiger, 1985) and to the 8th edition of *Grant's Atlas of Anatomy* (Baltimore, The Williams & Wilkins Co., 1983). The plate of the human skeleton, found in the previous editions, has been replaced by a schematic diagram of the brachial plexus (p. 112). The format of the book remains unchanged.

Diagrammatic representation and condensed description cannot do full justice to the complex relations involved. The illustrations do not attempt to show all details of attachments, nerves and arteries, since the object is to emphasize the major termini of muscles and the chief arteries and nerves related to them. Likewise the legends stress the primary functions, which imply movement at the insertions. The student must realize, however, that when the insertions are fixed, muscles produce movement at their origins as well. In the lower extremity this occurs almost as frequently as primary action.

Motor points were tested on normal subjects. Since the points vary among individuals, diagrams can give only the approximate location of greatest muscular response. Motor points are not included for muscles which do not show a clear-cut response to electrical stimulus.

A special word of thanks to Dr. Gerald C. Kraft, P.T. for reviewing the motor points. Four of these have been updated, consistent with current clinical findings. They include those for the trapezius, biceps brachii, brachioradialis and tibialis anterior muscles.

Buffalo, New York   John H. Warfel

5

CONTENTS

DORSAL MUSCLES OF THE FOREARM

MUSCLES OF THE HAND

MUSCLES OF THE LOWER EXTREMITY, ILIAC REGION

ANTERIOR MUSCLES OF THE THIGH

MEDIAL MUSCLES OF THE THIGH

MUSCLES OF THE GLUTEAL REGION

POSTERIOR MUSCLES OF THE THIGH

ANTERIOR MUSCLES OF THE LEG

POSTERIOR MUSCLES OF THE LEG

LATERAL MUSCLES OF THE LEG

MUSCLES OF THE FOOT

CHARTS

THE EXTREMITIES
Muscles and Motor Points

TRAPEZIUS

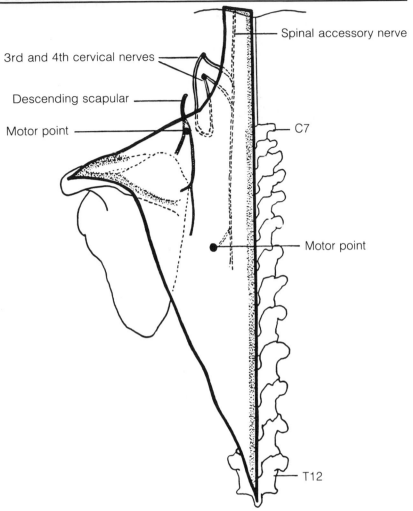

- Spinal accessory nerve
- 3rd and 4th cervical nerves
- Descending scapular
- Motor point
- C7
- Motor point
- T12

ORIGIN: External occipital protuberance, superior nuchal line, nuchal ligament from spines of seventh cervical and all thoracic vertebrae

INSERTION: Lateral third of clavicle, spine of scapula, acromion

FUNCTION: Adducts scapula, tilts chin, draws back acromion, rotates scapula

NERVE: Spinal Accessory, 3d and 4th cervical

ARTERY: Descending scapular (transverse cervical)

References

	GRAY	GRANT'S ATLAS
Muscle	513	5-25A, 6-30, 9-5
Nerve	513, 1189, 1199, 1200, 1205	5-25A, 6-30, 6-31, 8-11, 9-4
Artery	704, 706	6-31

LATISSIMUS DORSI

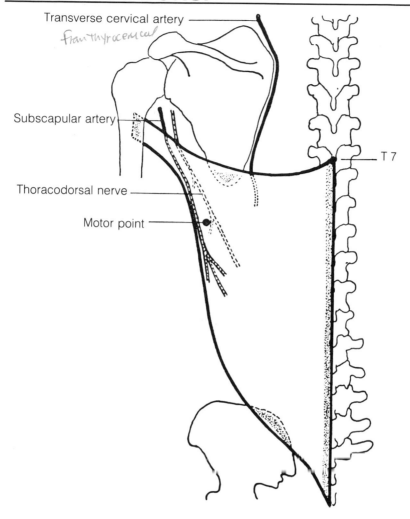

Transverse cervical artery

From thyrocervical

Subscapular artery

Thoracodorsal nerve

Motor point

T 7

ORIGIN: Spines of lower 6 thoracic vertebrae, lumbodorsal fascia, crest of ilium, muscular slips from lower 3 or 4 ribs

INSERTION: Floor of bicipital groove of humerus

FUNCTION: Adducts, extends, and medially rotates humerus

NERVE: Thoracodorsal

ARTERY: Descending scapular (transverse cervical), subscapular

References

	GRAY	GRANT'S ATLAS
Muscle	513	5-25A
Nerve	515, 1205, 1207, 1210	6-18, 6-22, 6-23
Artery	706, 714	6–22

13

RHOMBOIDEUS MAJOR

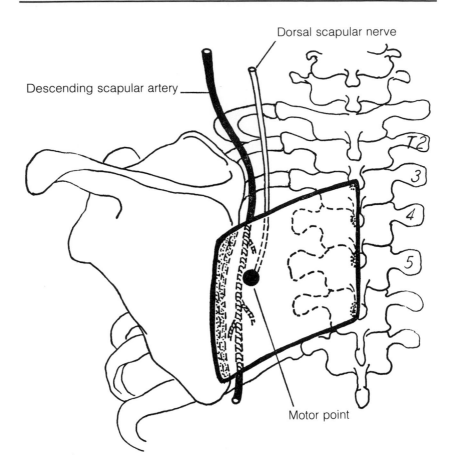

Dorsal scapular nerve

Descending scapular artery

T2

3

4

5

Motor point

ORIGIN: Spine of 2d, 3d, 4th and 5th thoracic vertebra
INSERTION: Medial border of scapula, between spine and inferior angle
FUNCTION: Adducts and rotates scapula
NERVE: Dorsal scapular
ARTERY: Descending scapular (transverse cervical)

References

	GRAY	GRANT'S ATLAS
Muscle	515	5-25A, 5-26
Nerve	516, 1205, 1207	9-5, 9-6
Artery	706	9-5, 9-6

RHOMBOIDEUS MINOR

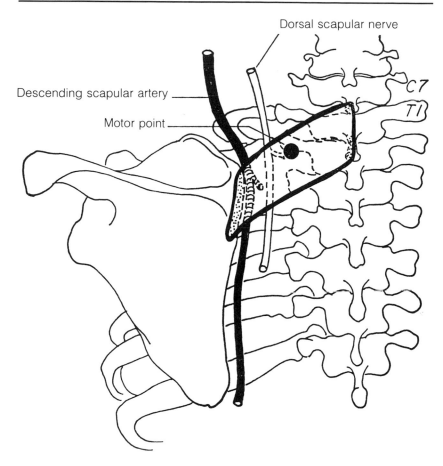

Dorsal scapular nerve

Descending scapular artery

Motor point

C7

T1

ORIGIN: Ligamentum nuchae, spine of 7th cervical and 1st thoracic vertebra

INSERTION; Root of scapular spine

FUNCTION: Adducts and rotates scapula

NERVE: Dorsal scapular

ARTERY: Descending scapular (transverse cervical)

References

	GRAY	GRANT'S ATLAS
Muscle	515	5-25A, 5-26
Nerve	516, 1205, 1207	9-5, 9-6
Artery	706	9-5, 9-6

LEVATOR SCAPULAE

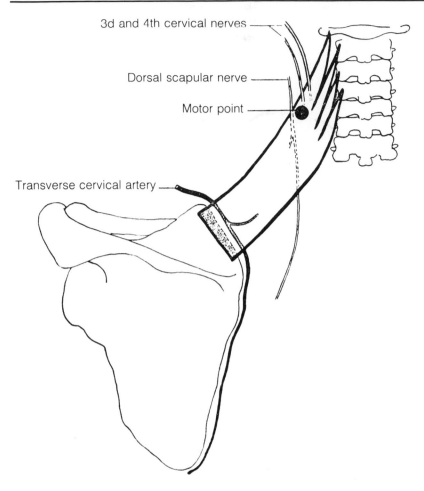

3d and 4th cervical nerves

Dorsal scapular nerve

Motor point

Transverse cervical artery

ORIGIN: Transverse process of atlas, axis, 3d and 4th cervical vertebra

INSERTION: Vertebral border of scapula between superior (medial) angle and root of spine

FUNCTION: Raises scapula or inclines neck to corresponding side if scapula is fixed

NERVE: Dorsal scapular, 3d and 4th cervical

ARTERY: Descending scapular (transverse cervical)

References

	GRAY	GRANT'S ATLAS
Muscle	516	5-25A, 5-26
Nerve	516, 1200, 1205, 1207	9-5, 9-6
Artery	706	9-6, 9-7

PECTORALIS MAJOR   *must superficial*
(-) platysma

Internal thoracic artery

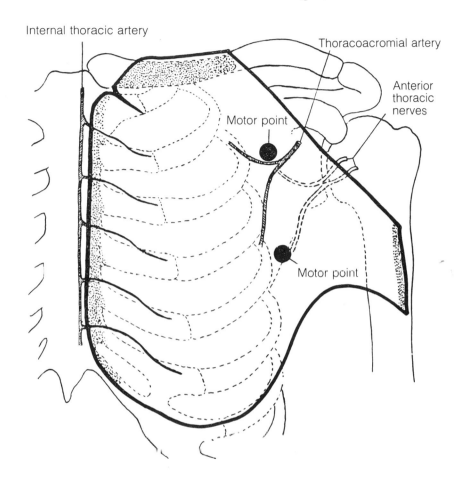

Thoracoacromial artery

Anterior thoracic nerves

Motor point

Motor point

ORIGIN: Sternal half of clavicle, sternum to 7th rib, cartilages of true ribs, aponeurosis of external oblique muscle
INSERTION: Lateral lip of bicipital groove of humerus   *or inter tubercular sulcus*
FUNCTION: Adducts arm, draws it forward, rotates it medially
NERVE: Medial and lateral anterior thoracic (medial and lateral pectoral)
ARTERY: Pectoral branch of thoracoacromial, perforating branches of internal thoracic

References

	GRAY	GRANT'S ATLAS
Muscle	518	6-14
Nerve	519, 1205, 1207, 1209	6-16, 6-20
Artery	707, 712, 713	6-16, 6-20

PECTORALIS MINOR

deep to pec. major

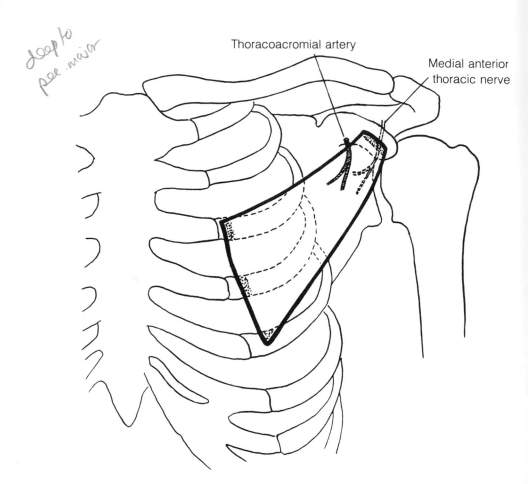

Thoracoacromial artery

Medial anterior thoracic nerve

ORIGIN: Outer surface of upper margin of 3d, 4th, and 5th rib
INSERTION: Coracoid process of scapula
FUNCTION: Lowers lateral angle of scapula, pulls shoulder forward
NERVE: Medial anterior thoracic (medial pectoral)
ARTERY: Thoracoacromial and intercostal branches of internal thoracic, lateral thoracic

References

	GRAY	GRANT'S ATLAS
Muscle	520	1-15
Nerve	520, 1205, 1207, 1209	6-20
Artery	707, 712, 713	6-20

SUBCLAVIUS

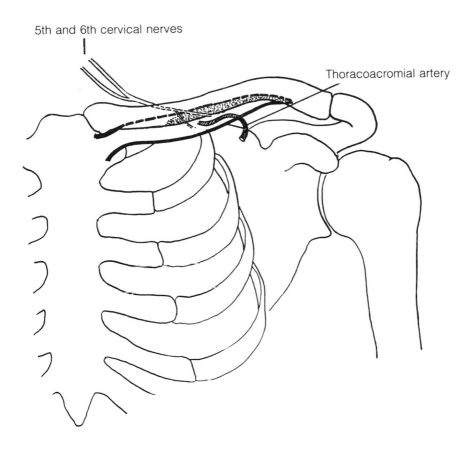

5th and 6th cervical nerves

Thoracoacromial artery

ORIGIN: Upper border of 1st rib and its cartilage
INSERTION.   Groove on under surface of clavicle
FUNCTION: Draws clavicle down and forward
NERVE: 5th and 6th cervical (nerve to subclavius)
ARTERY: Clavicular branch of thoracoacromial

References

	GRAY	GRANT'S ATLAS
Muscle	521	6-20, 9-7
Nerve	521, 1205, 1207, 1209	9-7
Artery	712	6-16

SERRATUS ANTERIOR

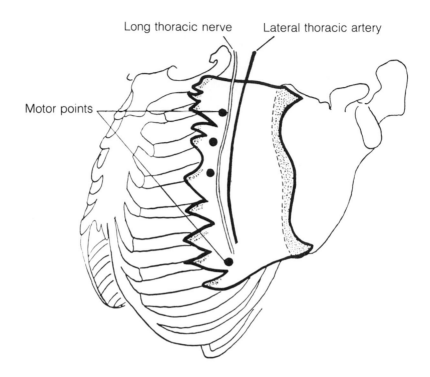

Long thoracic nerve   Lateral thoracic artery

Motor points

ORIGIN: Outer surface of upper 8 or 9 ribs
INSERTION: Costal surface of vertebral border of scapula
FUNCTION: Abducts scapula; raises ribs when scapula is fixed
NERVE: Long thoracic
ARTERY: Lateral thoracic

References

	GRAY	GRANT'S ATLAS
Muscle	521	5-26, 6-28
Nerve	521, 1205, 1207	2-5, 6-23, 6-28
Artery	713	6-20

DELTOIDEUS

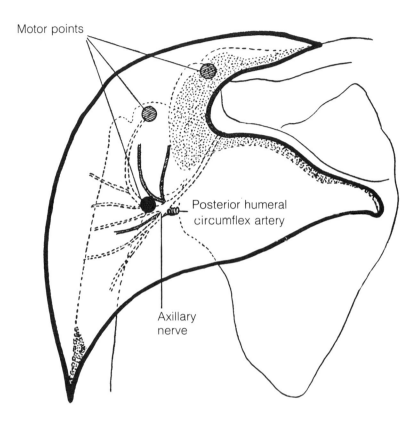

Motor points

Posterior humeral circumflex artery

Axillary nerve

ORIGIN: Lateral third of clavicle, upper surface of acromion, spine of scapula

INSERTION: Deltoid tuberosity of humerus

FUNCTION: Abducts arm

NERVE: Anterior and posterior branches of axillary (circumflex)

ARTERY: Posterior humeral circumflex, deltoid branch of thoracoacromial

References

	GRAY	GRANT'S ATLAS
Muscle	522	6-16, 6-37A
Nerve	522, 1205, 1207, 1210	6-24, 6-39
Artery	712, 714	6-39

SUBSCAPULARIS

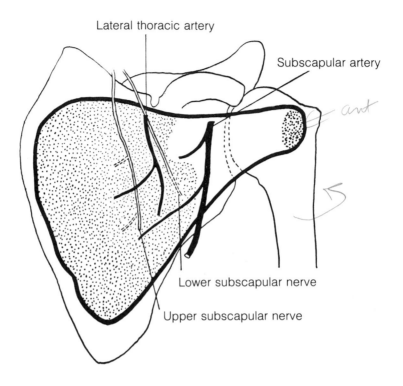

Lateral thoracic artery

Subscapular artery

Lower subscapular nerve

Upper subscapular nerve

ORIGIN: Subscapular fossa
INSERTION: Lesser tuberosity of humerus and capsule of shoulder joint
FUNCTION: Rotates humerus medially, draws it forward and down when arm
             is raised
NERVE: Upper and lower subscapular
ARTERY: Lateral thoracic, subscapular

References

	GRAY	GRANT'S ATLAS
Muscle	522	6-23, 6-48A
Nerve	523, 1205, 1207, 1209	6-18, 6-23
Artery	713, 714	6-20, 6-22

SUPRASPINATUS

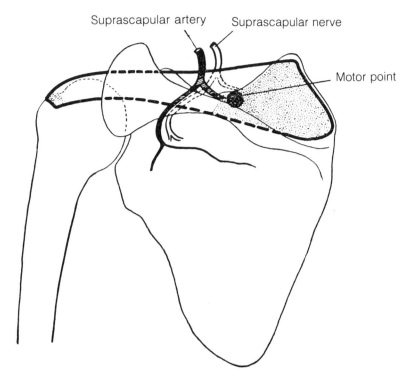

Suprascapular artery   Suprascapular nerve

Motor point

ORIGIN: Supraspinous fossa of scapula

INSERTION: Superior facet of greater tuberosity of humerus; capsule of shoulder joint

FUNCTION: Assists deltoid in abducting arm, fixes head of humerus in glenoid cavity; rotates head of humerus laterally

NERVE: Suprascapular

ARTERY: Suprascapular (transverse scapular)

References

	GRAY	GRANT'S ATLAS
Muscle	523	6-33, 6-48A
Nerve	523, 1205, 1207, 1209	6-31, 6-39
Artery	703	6-31

INFRASPINATUS

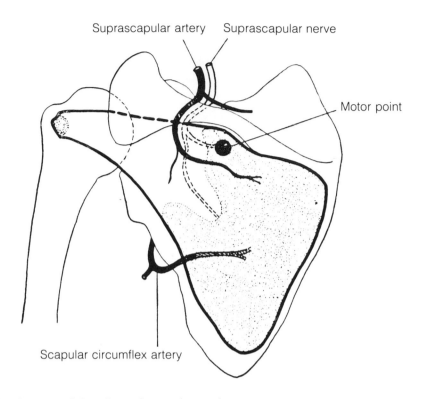

Suprascapular artery   Suprascapular nerve

Motor point

Scapular circumflex artery

ORIGIN: Infraspinous fossa of scapula
INSERTION: Middle facet of greater tuberosity of humerus; capsule of shoulder
                    joint
FUNCTION: Rotates head of humerus laterally with teres minor
NERVE: Suprascapular
ARTERY: Suprascapular (transverse scapular); scapular circumflex

References

	GRAY	GRANT'S ATLAS
Muscle	523	6-39, 6-48A
Nerve	524, 1205, 1207, 1209	6-31, 6-39
Artery	703, 713	6-39

TERES MINOR

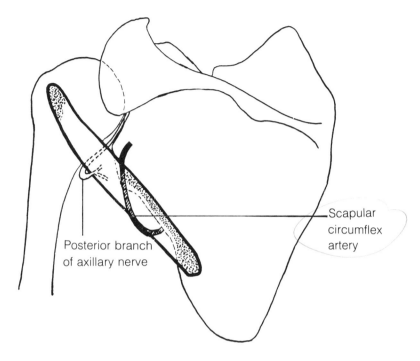

Posterior branch
of axillary nerve

Scapular
circumflex
artery

ORIGIN: Dorsal surface of axillary border of scapula

INSERTION. Lowest facet of greater tuberosity of humerus; capsule of shoulder joint

FUNCTION: Adducts and rotates head of humerus laterally and draws humerus toward glenoid fossa

NERVE: Posterior branch of axillary (circumflex)

ARTERY: Scapular circumflex

References

	GRAY	GRANT'S ATLAS
Muscle	524	6-39, 6-48A
Nerve	524, 1205, 1207, 1210	6-24, 6-39
Artery	713	6-39

TERES MAJOR

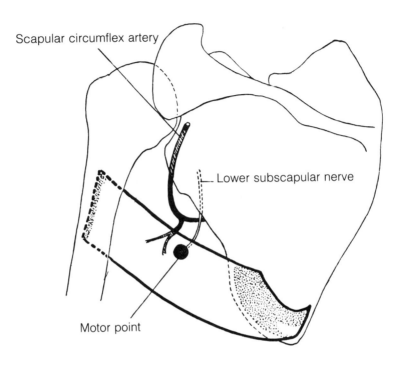

Scapular circumflex artery

Lower subscapular nerve

Motor point

ORIGIN: Dorsal surface of inferior angle of scapula
INSERTION: Medial lip of bicipital groove of humerus
FUNCTION: Adducts and medially rotates humerus and draws it back
NERVE: Lower subscapular
ARTERY: Scapular circumflex

References

	GRAY	GRANT'S ATLAS
Muscle	524	6-23, 6-39
Nerve	524, 1205, 1207, 1209	6-18, 6-23
Artery	713	6-39

CORACOBRACHIALIS

ANT (Flex) Compartment
3 muscle

Musculocutaneous nerve

Brachial artery

Musculocutaneous n.
brachial artery

ORIGIN: Tip of coracoid process of scapula
INSERTION: (Middle) of medial border of humerus
FUNCTION: Flexion and adduction of arm
NERVE: Musculocutaneous
ARTERY: Muscular branches of brachial

References

	GRAY	GRANT'S ATLAS
Muscle	526	6-20, 6-23
Nerve	526, 1205, 1207, 1212	6-23, 6-24, 6-25
Artery	719	6-25

BICEPS BRACHII

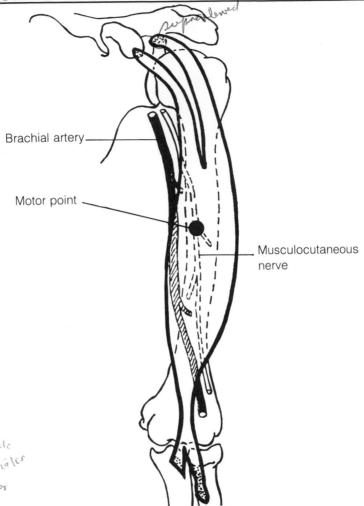

Brachial artery

Motor point

Musculocutaneous nerve

supraglenoid

coracoid O.
This muscle
corobrachialis
P. minor

ORIGIN: Short head from tip of coracoid process of scapula, long head from supraglenoid tuberosity of scapula

INSERTION: Radial tuberosity and by lacertus fibrosus to origins of forearm flexors

FUNCTION: Flexes and supinates forearm, flexes arm when forearm is fixed

NERVE: Musculocutaneous

ARTERY: Muscular branches of brachial

References

	GRAY	GRANT'S ATLAS
Muscle	527	6-18, 6-25, 6-43, 6-53
Nerve	528, 1205, 1207, 1212	6-23
Artery	719	6-23, 6-25

BRACHIALIS

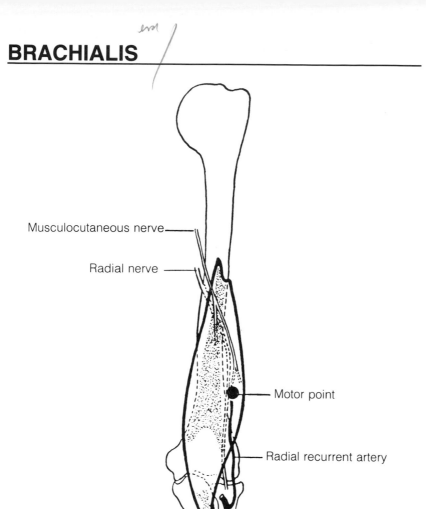

ORIGIN: Lower two-thirds of anterior surface of humerus
INSERTION: Coronoid process and tuberosity of ulna
FUNCTION: Flexes forearm
NERVE: Musculocutaneous, radial (may be afferent)
ARTERY: Radial recurrent, brachial

References

	GRAY	GRANT'S ATLAS
Muscle	528	6-37A, 6-55
Nerve	528, 1205, 1207, 1212, 1220	6-23
Artery	719, 726	6-25

TRICEPS BRACHII

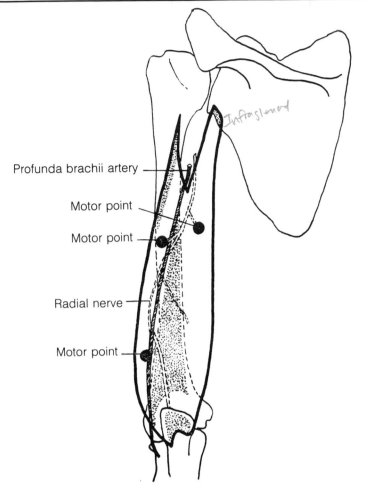

Profunda brachii artery

Infraglenoid

Motor point

Motor point

Radial nerve

Motor point

ORIGIN: Long head from infraglenoid tuberosity of scapula, lateral head from posterior and lateral surface of humerus, medial head from lower posterior surface of humerus

INSERTION: Upper posterior surface of olecranon and deep fascia of forearm

FUNCTION: Extends forearm; if arm is abducted, long head aids in adducting it

NERVE: Radial

ARTERY: Branch of profunda brachii

References

	GRAY	GRANT'S ATLAS
Muscle	528	6-25, 6-39, 6-62
Nerve	529, 1205, 1207, 1220	6-40B
Artery	718	6-23, 6-40B

30

PRONATOR TERES

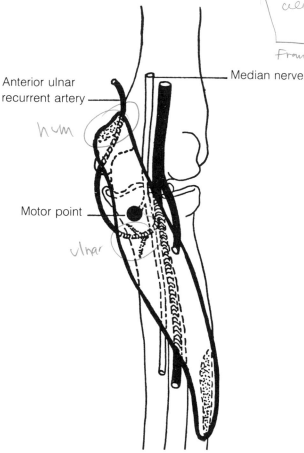

Anterior ulnar recurrent artery

Median nerve

Motor point

(handwritten: hum, ulnar)

ORIGIN: Humeral head from medial epicondylar ridge of humerus and common flexor tendon, ulnar head from medial side of coronoid process of ulna

INSERTION: Middle of lateral surface of radius

FUNCTION: Pronates forearm, assists in flexing forearm

NERVE: Median

ARTERY: Anterior ulnar recurrent

References

	GRAY	GRANT'S ATLAS
Muscle	530	6-66
Nerve	531, 1205, 1207, 1213	6-68
Artery	720, 726	Not shown

31

FLEXOR CARPI RADIALIS

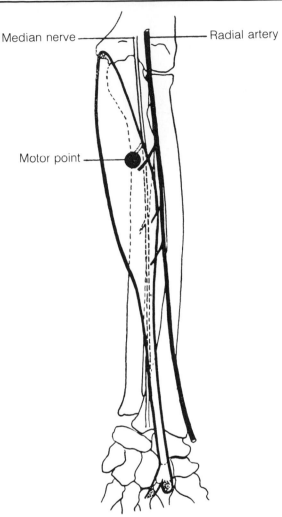

Median nerve — Radial artery

Motor point

ORIGIN: Common flexor tendon from medial epicondyle of humerus, fascia of forearm

INSERTION: Base of 2d and 3d metacarpal bones

FUNCTION: Flexes wrist, assists in pronating and abducting hand, assists in flexing forearm

NERVE: Median

ARTERY: Muscular branches of radial

References

	GRAY	GRANT'S ATLAS
Muscle	531	6-66
Nerve	531, 1205, 1207, 1213	Not shown
Artery	726	Not shown

PALMARIS LONGUS   Not always present

Sup

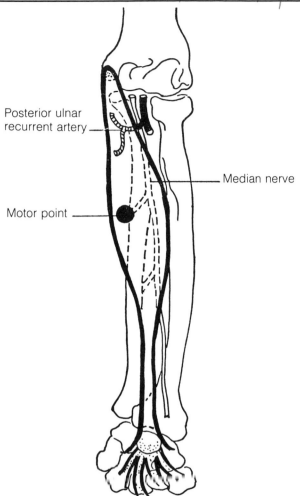

Posterior ulnar
recurrent artery

Median nerve

Motor point

ORIGIN: Common flexor tendon from medial epicondyle of humerus
INSERTION: Transverse carpal ligament and palmar aponeurosis
FUNCTION: Flexes wrist, assists in pronation and flexion of forearm
NERVE: Median
ARTERY: Posterior ulnar recurrent

References

	GRAY	GRANT'S ATLAS
Muscle	531	6-66, 6-74
Nerve	532, 1205, 1207, 1213	Not shown
Artery	720	Not shown

FLEXOR CARPI ULNARIS

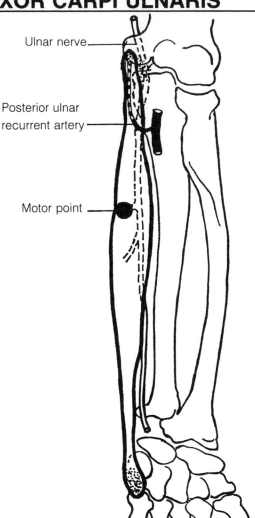

Ulnar nerve

Posterior ulnar recurrent artery

Motor point

Sup

painic

Ex. carpi ulnous

ORIGIN: <u>Humeral head</u> from common flexor tendon from medial epicondyle of humerus, <u>ulnar head</u> from olecranon and dorsal border
of ulna

INSERTION: Pisiform, hamate, 5th metacarpal bones

FUNCTION: Flexes wrist and assists in adducting it; assists in flexing forearm

NERVE: Ulnar

ARTERY: Posterior ulnar recurrent

References

	GRAY	GRANT'S ATLAS
Muscle	532	6-66, 6-68
Nerve	532, 1205, 1207, 1217	6-68, 6-69
Artery	720	6-64, 6-69

FLEXOR DIGITORUM SUPERFICIALIS

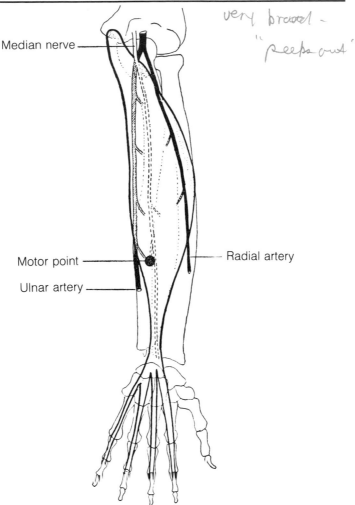

very broad -

"peeks out"

Median nerve

Motor point

Ulnar artery

Radial artery

ORIGIN: Humeral head from common flexor tendon from medial epicondyle of humerus, ulnar head from coronoid process of ulna, radial
head from oblique line of radius

INSERTION: Margins of palmar surface of middle phalanx of medial 4 digits *2-5*
button holes
for deep
flexor
tendons

FUNCTION: Flexes middle and proximal phalanges of medial 4 digits, aids in
flexing wrist and forearm

NERVE: Median

ARTERY: Muscular branches of ulnar, muscular branches of radial

References

	GRAY	GRANT'S ATLAS
Muscle	532	6-66, 6-74, 6-85A
Nerve	533, 1205, 1207, 1213	Not shown
Artery	721, 726	6-68

FLEXOR DIGITORUM PROFUNDUS ①

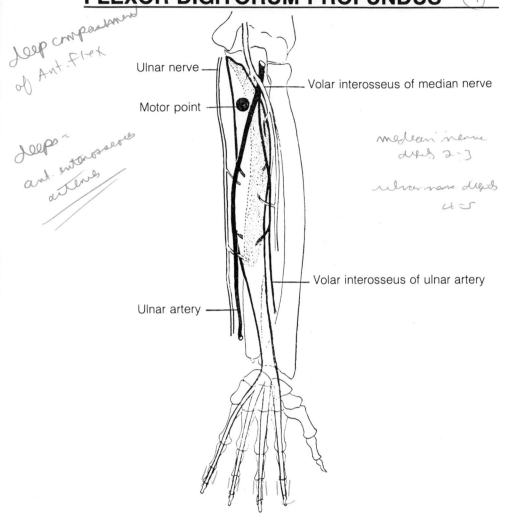

deep compartment of Ant. Flex

deeps — ant. interosseus artene

Ulnar nerve

Motor point

Volar interosseus of median nerve

median nerve dig's 2-3

ulnar nerve dig's 4-5

Volar interosseus of ulnar artery

Ulnar artery

ORIGIN: Medial and anterior surface of ulna, interosseus membrane, deep fascia of forearm

INSERTION: Distal phalanges of medial 4 digits

FUNCTION: Flexes terminal phalanges of medial 4 digits after superficialis flexes 2d phalanges, aids in flexing wrist

NERVE: Ulnar, volar interosseus of median

ARTERY: Volar interosseus of ulnar, muscular branches of ulnar

References

	GRAY	GRANT'S ATLAS
Muscle	533	6-69, 6-85A
Nerve	534, 1205, 1207, 1213, 1217	6-68, 6-69
Artery	721	6-69

36

FLEXOR POLLICIS LONGUS

jeep

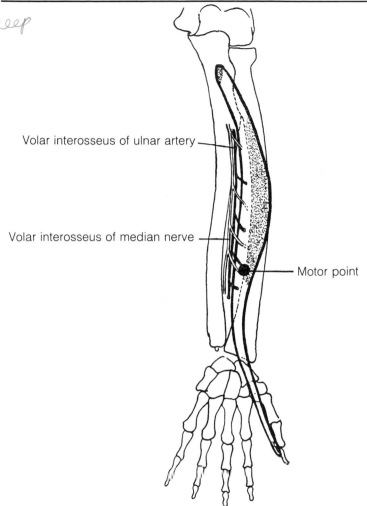

Volar interosseus of ulnar artery

Volar interosseus of median nerve

Motor point

ORIGIN: Volar surface of radius, adjacent interosseus membrane, medial border of coronoid process of ulna

INSERTION: Base of distal phalanx of thumb on palmar surface

FUNCTION: Flexes thumb

NERVE: Volar interosseus of median

ARTERY: Volar interosseus of ulnar

References

	GRAY	GRANT'S ATLAS
Muscle	535	6-69, 6-83
Nerve	535, 1205, 1207, 1213	6-69
Artery	721	6-69

PRONATOR QUADRATUS

deep

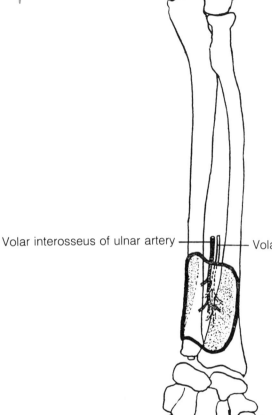

Volar interosseus of ulnar artery ———— Volar interosseus of median nerve

ORIGIN: Distal fourth of volar surface of ulna
INSERTION: Distal fourth of lateral border on volar surface of radius
FUNCTION: Pronates forearm
NERVE: Volar interosseus of median
ARTERY: Volar interosseus of ulnar

References

	GRAY	GRANT'S ATLAS
Muscle	535	6-85A
Nerve	535, 1205, 1207, 1213	Not shown
Artery	721	Not shown

BRACHIORADIALIS

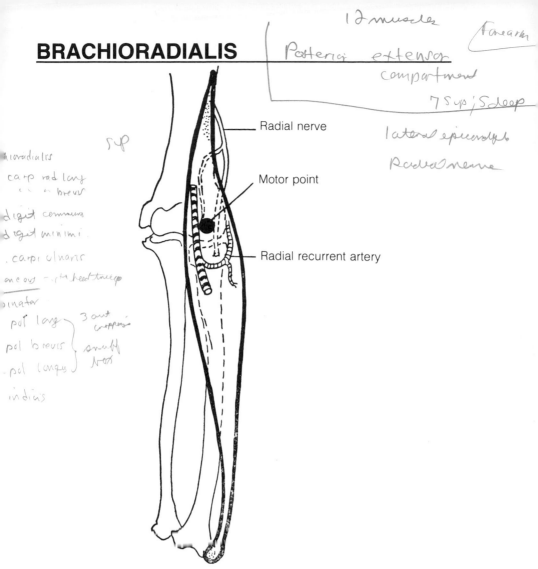

Handwritten annotations:
- 12 muscles
- Posterior extensor compartment (forearm)
- 7 sup; 5 deep
- lateral epicondyle
- Radial nerve
- brachioradialis
- carp rad long
- " " brevis
- digit communis
- digit minimi
- carpi ulnaris
- aneous - 1st head triceps
- inator
- pol long ┐ 3 out
- pol brevis │ snuff
- pol longus ┘ box
- indicis
- sup

Diagram labels:
- Radial nerve
- Motor point
- Radial recurrent artery

ORIGIN: Proximal two-thirds of lateral supracondylar ridge of humerus, lateral intermuscular septum

INSERTION: Lateral side of base of styloid process of radius

FUNCTION: Flexes forearm after flexion has been started by biceps and brachialis; may also act as a semipronator and semisupinator

NERVE: Radial

ARTERY: Radial recurrent

References

	GRAY	GRANT'S ATLAS
Muscle	535	6-66
Nerve	536, 1205, 1207, 1220	6-69
Artery	726	6-68, 6-69

EXTENSOR CARPI RADIALIS LONGUS

Sup

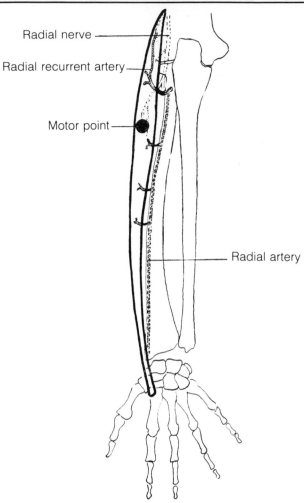

Radial nerve

Radial recurrent artery

Motor point

Radial artery

ORIGIN: Lower third of lateral supracondylar ridge of humerus, lateral intermuscular septum

INSERTION: Dorsal surface of base of 2d metacarpal bone

FUNCTION: Extends wrist, abducts hand

NERVE: Radial

ARTERY: Muscular branches of radial, radial recurrent

References

	GRAY	GRANT'S ATLAS
Muscle	536	6-91A, 6-104
Nerve	536, 1205, 1207, 1220	Not shown
Artery	726	6-69

40

EXTENSOR CARPI RADIALIS BREVIS

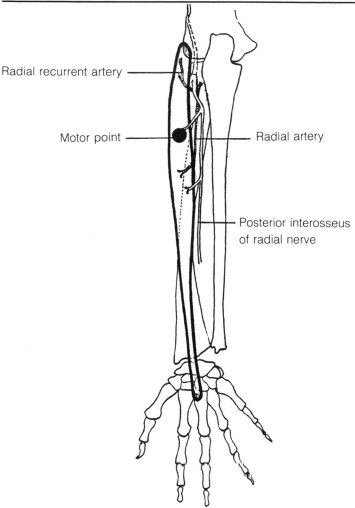

Radial recurrent artery

Motor point

Radial artery

Posterior interosseus
of radial nerve

ORIGIN: From common extensor tendon from lateral epicondyle of humerus, radial collateral ligament of elbow joint, intermuscular
septa

INSERTION: Dorsal surface of base of 3d metacarpal bone

FUNCTION: Extends wrist, abducts hand

NERVE: Posterior interosseus of radial

ARTERY: Muscular branches of radial, radial recurrent

References

	GRAY	GRANT'S ATLAS
Muscle	536	6-91A, 6-104
Nerve	537, 1205, 1207, 1221	6-69
Artery	726	6-69

EXTENSOR DIGITORUM COMMUNIS

See

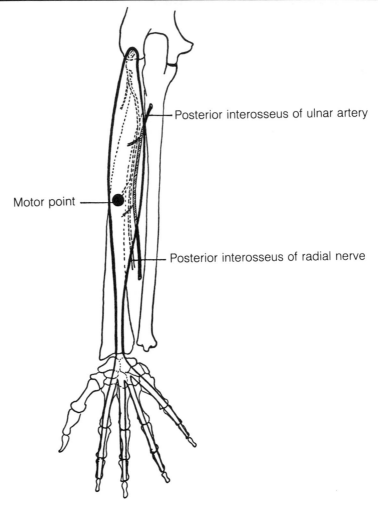

Posterior interosseus of ulnar artery

Motor point

Posterior interosseus of radial nerve

ORIGIN: Lateral epicondyle of humerus by common extensor tendon, intermuscular septa

proximid, distal phalanges

INSERTION: Lateral and dorsal surface of phalanges of medial 4 digits via extensor expansion

FUNCTION: Extends medial 4 digits; assists in extension of wrist

NERVE: Posterior interosseus of radial

ARTERY: Posterior interosseus of ulnar

References

	GRAY	GRANT'S ATLAS
Muscle	537	6-91A, 6-104
Nerve	537, 1205, 1207, 1221	6-93
Artery	721	6-93

EXTENSOR DIGITI MINIMI

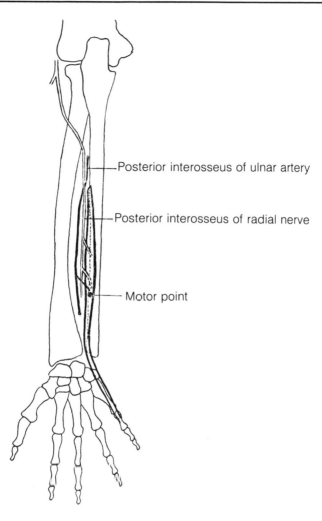

Posterior interosseus of ulnar artery

Posterior interosseus of radial nerve

Motor point

ORIGIN: Common extensor tendon from lateral epicondyle of the humerus, intermuscular septa

INSERTION: Dorsum of proximal phalanx of 5th digit

FUNCTION: Extends 5th digit

NERVE: Posterior interosseus of radial

ARTERY: Posterior interosseus of ulnar

References

	GRAY	GRANT'S ATLAS
Muscle	537	6-91A, 6-104
Nerve	538, 1205, 1207, 1221	6-93
Artery	721	6-93

EXTENSOR CARPI ULNARIS

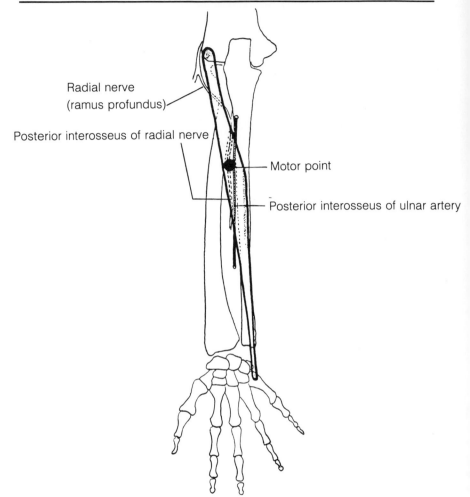

Radial nerve
(ramus profundus)

Posterior interosseus of radial nerve

Motor point

Posterior interosseus of ulnar artery

ORIGIN: From common extensor tendon from lateral epicondyle of
             humerus, and from posterior border of ulna
INSERTION: Medial side of base of 5th metacarpal bone
FUNCTION: Extends wrist, adducts hand
NERVE: Posterior interosseus of radial
ARTERY: Posterior interosseus of ulnar

References

	GRAY	GRANT'S ATLAS
Muscle	538	6-91A
Nerve	538, 1205, 1207, 1221	6-93
Artery	721	6-93

ANCONEUS

Sup

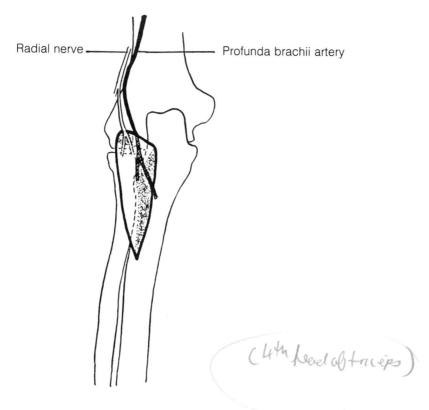

Radial nerve ———————— Profunda brachii artery

(4th level of triceps)

ORIGIN: Lateral epicondyle of humerus, posterior ligament of elbow joint
INSERTION: Lateral side of olecranon and posterior surface of ulna
FUNCTION: Assists triceps in extending forearm
NERVE: Radial
ARTERY: Branch of profunda brachii

References

	GRAY	GRANT'S ATLAS
Muscle	538	6-92
Nerve	538, 1205, 1207, 1220	6-91A
Artery	717	Not shown

SUPINATOR

Deep

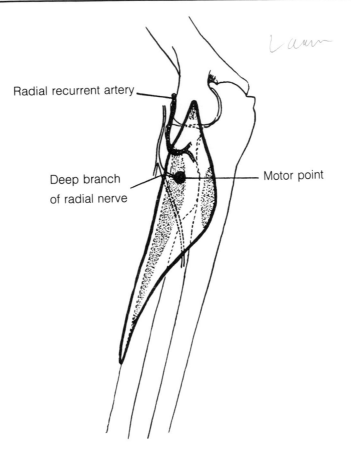

Radial recurrent artery

Deep branch
of radial nerve

Motor point

ORIGIN: Lateral epicondyle of humerus, lateral ligament of elbow joint and annular ligament of radius, supinator crest and fossa of ulna

INSERTION: Lateral and anterior surface of radius in its upper third

FUNCTION: Supinates forearm

NERVE: Deep branch of radial

ARTERY: Radial recurrent, posterior interosseous of ulnar

References

	GRAY	GRANT'S ATLAS
Muscle	538	6-55, 6-93
Nerve	539, 1205, 1207, 1221	6-69
Artery	721, 726	6-69

ABDUCTOR POLLICIS LONGUS

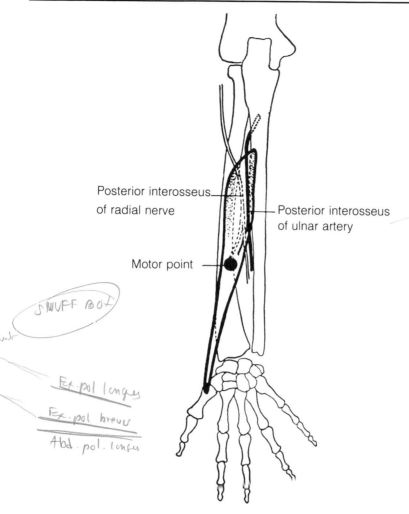

Posterior interosseus of radial nerve

Posterior interosseus of ulnar artery

Motor point

SNUFF BOT

Ex. pol longus
Ex. pol breuv
Abd. pol. longus

ORIGIN: Posterior surface of ulna, interosseus membrane, middle third of posterior surface of radius
INSERTION: Radial side of base of 1st metacarpal bone
FUNCTION: Abducts thumb and wrist
NERVE: Posterior interosseus of radial
ARTERY: Posterior interosseus of ulnar

References

	GRAY	GRANT'S ATLAS
Muscle	539	6-78, 6-91A, 6-93
Nerve	539, 1205, 1207, 1221	6-93
Artery	721	6-93

EXTENSOR POLLICIS BREVIS

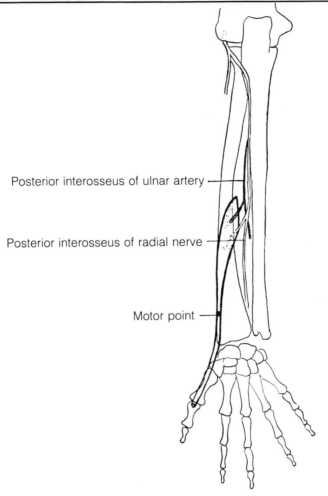

Posterior interosseus of ulnar artery —

Posterior interosseus of radial nerve —

Motor point —

ORIGIN: Posterior surface of radius, interosseus membrane
INSERTION: Base of proximal phalanx of thumb
FUNCTION: Extends proximal phalanx of thumb
NERVE: Posterior interosseus of radial
ARTERY: Posterior interosseus of ulnar

References

	GRAY	GRANT'S ATLAS
Muscle	540	6-91A, 6-93
Nerve	540, 1205, 1207, 1221	6-93
Artery	721	6-93

EXTENSOR POLLICIS LONGUS

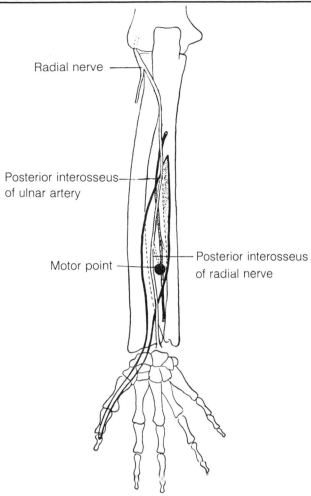

Radial nerve

Posterior interosseus of ulnar artery

Motor point

Posterior interosseus of radial nerve

ORIGIN: Middle third of posterior surface of ulna, interosseus membrane
INSERTION: Base of distal phalanx of thumb
FUNCTION: Extends terminal phalanx of thumb
NERVE: Posterior interosseus of radial
ARTERY: Posterior interosseus of ulnar

References

	GRAY	GRANT'S ATLAS
Muscle	540	6-91A, 6-93
Nerve	540, 1205, 1207, 1221	6-93
Artery	721	6-93

EXTENSOR INDICIS

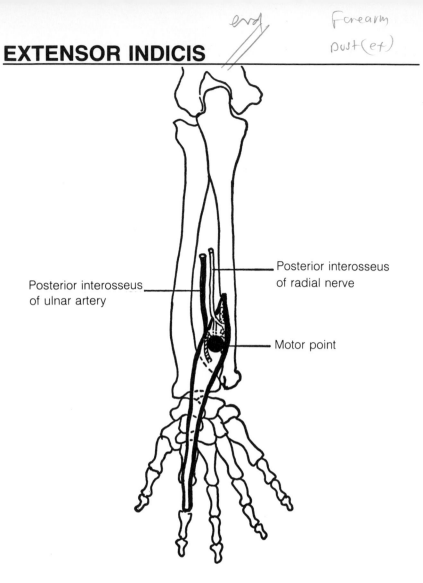

Posterior interosseus of radial nerve

Posterior interosseus of ulnar artery

Motor point

ORIGIN: Posterior surface of ulna, interosseus membrane
INSERTION: Dorsum of proximal phalanx of index finger
FUNCTION: Extends proximal phalanx of index finger
NERVE: Posterior interosseus of radial
ARTERY: Posterior interosseus of ulnar

References

	GRAY	GRANT'S ATLAS
Muscle	540	6-93, 6-104
Nerve	540, 1205, 1207, 1221	6-93
Artery	721	6-93

ABDUCTOR POLLICIS BREVIS

Thenar compartment

Median n.
Radial a.

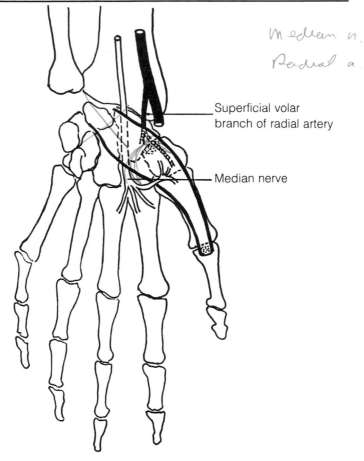

Superficial volar
branch of radial artery

Median nerve

ORIGIN: Transverse carpal ligament, scaphoid and trapezium bones
INSERTION: Radial side of base of proximal phalanx of thumb
FUNCTION: Abducts thumb, draws thumb forward at right angles to palm
NERVE: Muscular branches of median
ARTERY: Superficial volar branch of radial

References

	GRAY	GRANT'S ATLAS
Muscle	550	6-78
Nerve	550, 1205, 1207, 1216	6-78
Artery	727	6-78

OPPONENS POLLICIS

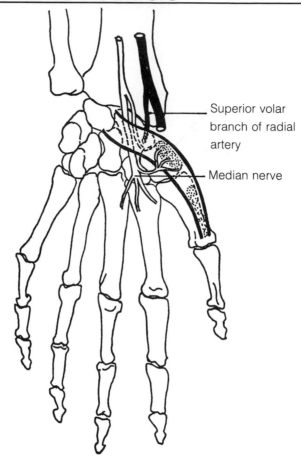

Superior volar branch of radial artery

Median nerve

ORIGIN: Transverse carpal ligament, trapezium bone

INSERTION: Anterior surface, radial side of 1st metacarpal bone

FUNCTION: Draws 1st metacarpal bone forward, and medially, opposing thumb to each of the other digits

NERVE: Muscular branches of median

ARTERY: Superficial volar branch of radial

References

	GRAY	GRANT'S ATLAS
Muscle	550	6-69, 6-79
Nerve	550, 1205, 1207, 1216	6-79
Artery	727	6-70

FLEXOR POLLICIS BREVIS

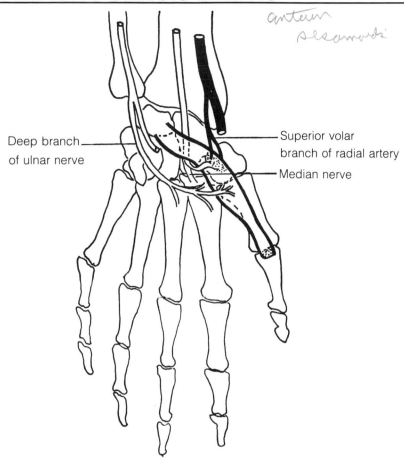

Deep branch of ulnar nerve

Superior volar branch of radial artery

Median nerve

ORIGIN: Transverse carpal ligament, trapezium bone
INSERTION: Base of proximal phalanx of thumb
FUNCTION: Flexes proximal phalanx of thumb
NERVE: Muscular branches of median; deep branch of ulnar
ARTERY: Superficial volar branch of radial

References

	GRAY	GRANT'S ATLAS
Muscle	550	6-69, 6-79
Nerve	551, 1205, 1207, 1216, 1219	6-79
Artery	727	6-78

ADDUCTOR POLLICIS

Adductor compartment
one muscle

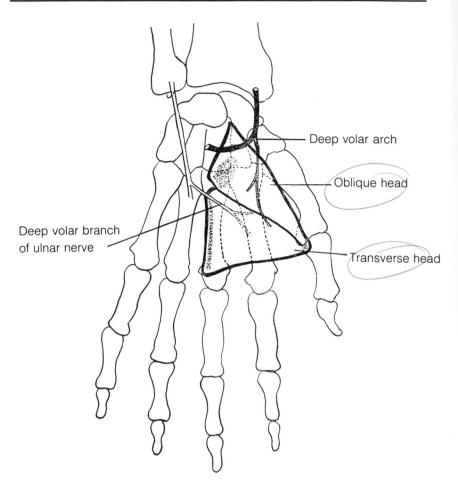

Deep volar arch

Oblique head

Deep volar branch
of ulnar nerve

Transverse head

ORIGIN: <u>Oblique head</u> from trapezium, trapezoid and capitate bones, base of 2d and 3d metacarpal bone, <u>transverse head</u> from volar surface of 3d metacarpal bone

INSERTION: Ulnar side of base of proximal phalanx of thumb

FUNCTION: Adducts thumb, aids in opposition

NERVE: Deep volar branch of ulnar

ARTERY: Deep volar arch

References

	GRAY	GRANT'S ATLAS
Muscle	551	6-76, 6-83, 6-87
Nerve	552, 1205, 1207, 1219	6-83
Artery	728	6-83

PALMARIS BREVIS *occasional muscle*

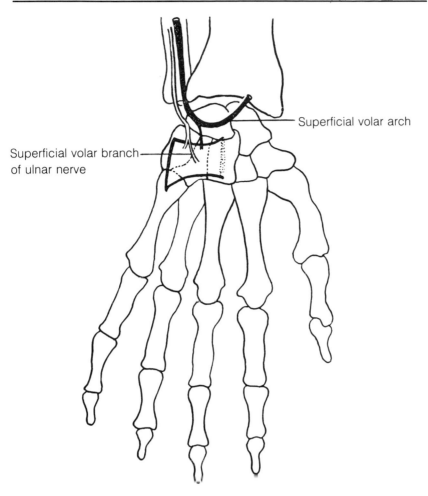

Superficial volar arch

Superficial volar branch of ulnar nerve

ORIGIN: Ulnar side of transverse carpal ligament, palmar aponeurosis

INSERTION: Skin on ulnar border of palm

FUNCTION: Corrugates skin on ulnar side of palm, deepens the hollow of the hand

NERVE: Superficial volar branch of ulnar

ARTERY: Superficial volar arch

References

	GRAY	GRANT'S ATLAS
Muscle	552	6-78
Nerve	552, 1205, 1207, 1218	Not shown
Artery	727	Not shown

ABDUCTOR DIGITI MINIMI

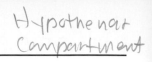

Hypothenar
Compartment

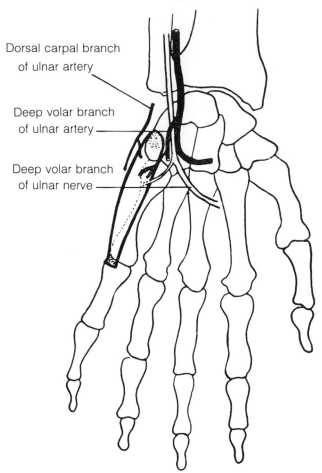

Dorsal carpal branch
of ulnar artery

Deep volar branch
of ulnar artery

Deep volar branch
of ulnar nerve

ORIGIN: Pisiform bone, tendon of flexor carpi ulnaris

INSERTION: Medial side of base of proximal phalanx of 5th digit and aponeurosis of extensor digiti minimi

FUNCTION: Abducts 5th digit from 4th digit

NERVE: Deep volar branch of ulnar

ARTERY: Deep volar branch of ulnar, dorsal carpal branch of ulnar

References

	GRAY	GRANT'S ATLAS
Muscle	552	6-78, 6-95
Nerve	553, 1205, 1207, 1218	6-79
Artery	722	6-83, 6-95

FLEXOR DIGITI MINIMI BREVIS

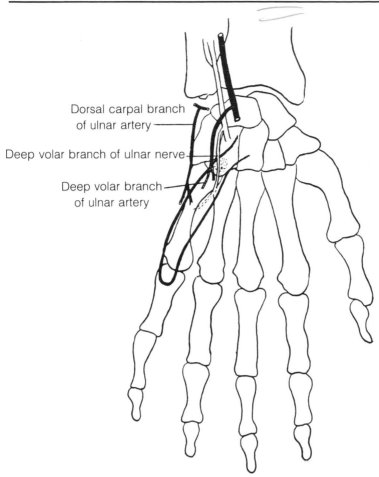

Dorsal carpal branch of ulnar artery —

Deep volar branch of ulnar nerve —

Deep volar branch of ulnar artery —

ORIGIN: Transverse carpal ligament, hamulus of hamate bone
INSERTION: Ulnar side, base of proximal phalanx of 5th digit
FUNCTION: Flexes proximal phalanx of 5th digit
NERVE: Deep volar branch of ulnar
ARTERY: Deep volar branch of ulnar, dorsal carpal branch of ulnar

References

	GRAY	GRANT'S ATLAS
Muscle	553	6-79
Nerve	554, 1205, 1207, 1218	6-79
Artery	722	6-83

OPPONENS DIGITI MINIMI

Opposition

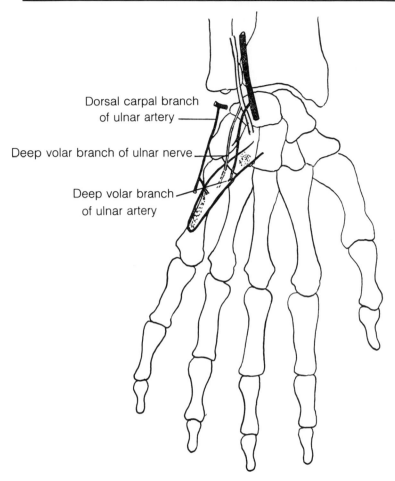

Dorsal carpal branch
of ulnar artery

Deep volar branch of ulnar nerve

Deep volar branch
of ulnar artery

ORIGIN: Transverse carpal ligament, hamulus of hamate bone

INSERTION: Ulnar margin of 5th metacarpal bone

FUNCTION: Draws 5th metacarpal bone forward to face thumb, deepens hollow of hand

NERVE: Deep volar branch of ulnar

ARTERY: Deep volar branch of ulnar, dorsal carpal branch of ulnar

References

	GRAY	GRANT'S ATLAS
Muscle	554	6-69, 6-79, 6-95
Nerve	554, 1205, 1207, 1218	6-79
Artery	722	6-83, 6-95

58

LUMBRICALES

MIDPALMAR COMPARTMENT

4 evi munle

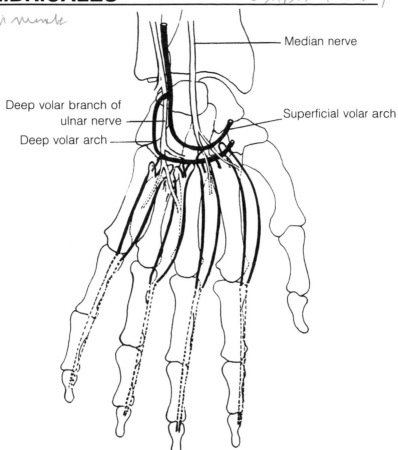

Median nerve

Deep volar branch of ulnar nerve

Deep volar arch

Superficial volar arch

ORIGIN: There are 4 lumbricales, all arising from tendons of flexor digitorum profundus: 1st from radial side of tendon for index finger, 2d from radial side of tendon for middle finger, 3d from adjacent sides of tendons for middle and ring fingers, 4th from adjacent sides of tendons for ring and little fingers

INSERTION: With tendons of extensor digitorum and interossei into bases of terminal phalanges of medial 4 digits

FUNCTION: Flex fingers at metacarpophalangeal joints, extend fingers at interphalangeal joints

NERVE: Median, to lateral two muscles, deep volar branch of ulnar to medial two muscles

ARTERY: Superficial and deep volar arches

References

	GRAY	GRANT'S ATLAS
Muscle	554	6-69, 6-79
Nerve	554, 1205, 1207, 1216, 1218	6-79
Artery	723, 728	6-78

INTEROSSEI DORSALES

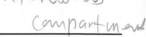

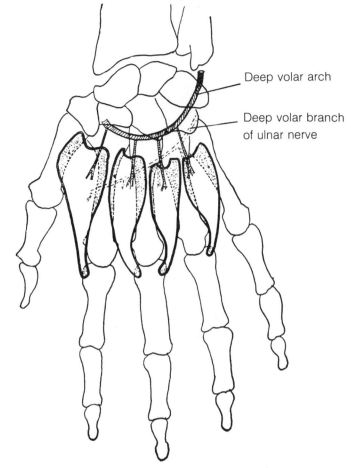

Deep volar arch

Deep volar branch
of ulnar nerve

ORIGIN: There are 4 dorsal interossei; each arises by 2 heads from adjacent sides of metacarpal bones

INSERTION: 1st into radial side of proximal phalanx of 2d digit, 2d into radial side of proximal phalanx of 3d digit, 3d into ulnar side of proximal phalanx of 3d digit, 4th into ulnar side of proximal phalanx of 4th digit

FUNCTION: Abduct index, middle and ring fingers from the mid line of the hand

NERVE: Deep volar branch of ulnar

ARTERY: Deep volar arch

References

	GRAY	GRANT'S ATLAS
Muscle	554	6-85A, 6-91A, 6-104
Nerve	555, 1205, 1207, 1218	6-85A
Artery	728	6-84, 6-88

INTEROSSEI PALMARES

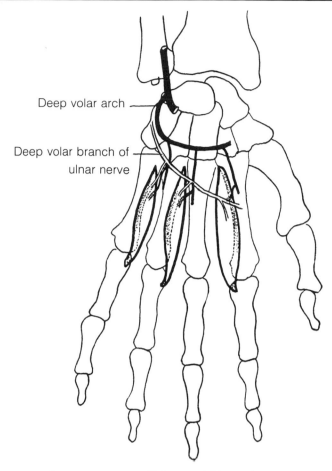

Deep volar arch

Deep volar branch of ulnar nerve

ORIGIN: There are 3 volar interossei. 1st from ulnar side of 2d metacarpal bone, 2d from radial side of 4th metacarpal bone, 3d from radial side of 5th metacarpal bone

INSERTION: 1st into ulnar side of proximal phalanx of 2d digit, 2d into radial side of proximal phalanx of 4th digit, 3d into radial side of proximal phalanx of 5th digit

FUNCTION: Each muscle adducts digit into which it is inserted toward middle digit

NERVE: Deep volar branch of ulnar

ARTERY: Deep volar arch

References

	GRAY	GRANT'S ATLAS
Muscle	555	6-85A
Nerve	555, 1205, 1207, 1218	6-85A
Artery	728	6-84

PSOAS MAJOR

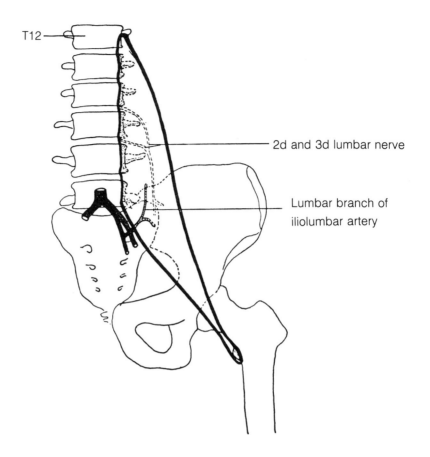

T12

2d and 3d lumbar nerve

Lumbar branch of
iliolumbar artery

ORIGIN: Anterior surface of transverse processes and bodies of lumbar
               vertebrae, corresponding intervertebral discs
INSERTION: Lesser trochanter of femur (with iliacus forms iliopsoas)
FUNCTION: Flexes thigh, flexes vertebral column on pelvis when leg is fixed
NERVE: 2d and 3d lumbar
ARTERY: Lumbar branch of iliolumbar

References

	GRAY	GRANT'S ATLAS
Muscle	557	2-99, 2-119, 4-17
Nerve	557, 1226	Not shown
Artery	756	Not shown

PSOAS MINOR

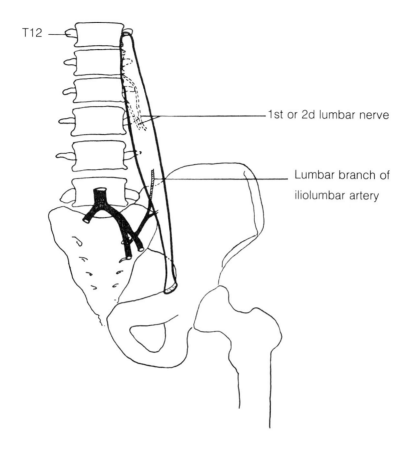

T12

1st or 2d lumbar nerve

Lumbar branch of
iliolumbar artery

ORIGIN: Vertebral margins of 12th thoracic and 1st lumbar vertebra, cor-
responding intervertebral disc
INSERTION: Pectineal line, iliopectineal eminence
FUNCTION: Flexes pelvis on vertebral column, assists psoas major in flexing
vertebral column on pelvis. This muscle is inconstant, absent in
40 per cent of bodies
NERVE: 1st or 2d lumbar
ARTERY: Lumbar branch of iliolumbar

References

	GRAY	GRANT'S ATLAS
Muscle	557	4-28A
Nerve	558, 1226	Not shown
Artery	756	Not shown

ILIACUS

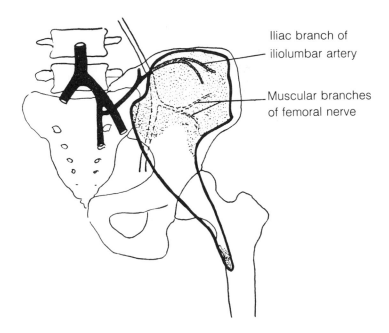

Iliac branch of
iliolumbar artery

Muscular branches
of femoral nerve

ORIGIN: Upper two-thirds of iliac fossa; iliac crest; anterior sacroiliac, lumbosacral, and iliolumbar ligaments; ala of sacrum

INSERTION: Tendon of psoas major, lesser trochanter, capsule of hip joint, body of femur (with psoas major forms iliopsoas)

FUNCTION: Flexes thigh, tilts pelvis forward when leg is fixed

NERVE: Muscular branches of femoral

ARTERY: Iliac branch of iliolumbar, superior gluteal

References

	GRAY	GRANT'S ATLAS
Muscle	558	4-28B
Nerve	558, 1226, 1232	2-119
Artery	756, 757	2-119

SARTORIUS

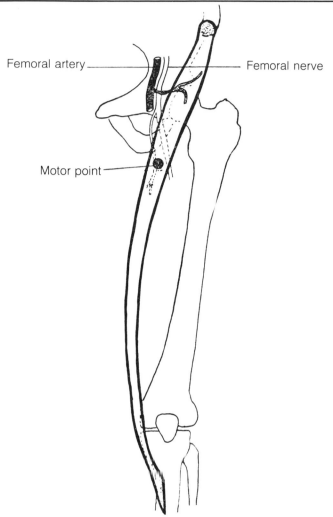

Femoral artery — Femoral nerve

Motor point —

ORIGIN: Anterior superior iliac spine, upper half of iliac notch
INSERTION: Upper part of medial surface of tibia
FUNCTION: Flexes leg on thigh, flexes thigh on pelvis, rotates thigh laterally
NERVE: Muscular branches of femoral
ARTERY: Muscular branches of femoral

References

	GRAY	GRANT'S ATLAS
Muscle	561	4-20, 4-28A
Nerve	562, 1226, 1232	4-25
Artery	765	Not shown

RECTUS FEMORIS

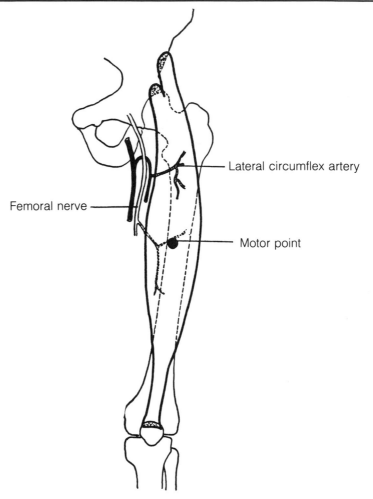

Lateral circumflex artery

Femoral nerve

Motor point

ORIGIN: (Rectus femoris is 1 of 4 muscles comprising quadriceps femoris.) Straight head from anterior inferior iliac spine, reflected head from groove on upper brim of acetabulum

INSERTION: Upper border of patella; by ligamentum patellae into tibial tuberosity

FUNCTION: Extends leg, flexes thigh

NERVE: Muscular branches of femoral

ARTERY: Lateral femoral circumflex

References

	GRAY	GRANT'S ATLAS
Muscle	**562-563**	**4-28A**
Nerve	**563, 1226, 1233**	**4-25**
Artery	**767**	**4-25**

VASTUS LATERALIS

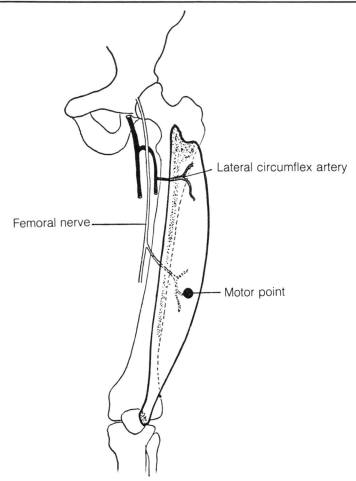

Lateral circumflex artery

Femoral nerve

Motor point

ORIGIN: (Vastus lateralis is 1 of 4 muscles comprising quadriceps femoris.) Capsule of hip joint, intertrochanteric line, greater trochanter, gluteal tuberosity, linea aspera, lateral intermuscular septum

INSERTION: Lateral border by patella, by ligamentum patellae into tibial tuberosity

FUNCTION: Extends leg

NERVE: Muscular branches of femoral

ARTERY: Lateral femoral circumflex, lateral superior genicular

References

	GRAY	GRANT'S ATLAS
Muscle	562-563	4-28A
Nerve	563, 1226, 1233	4-25
Artery	767, 772	4-25

VASTUS MEDIALIS

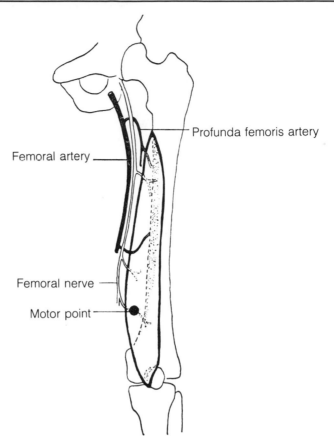

Profunda femoris artery

Femoral artery

Femoral nerve

Motor point

ORIGIN: (Vastus medialis is 1 of 4 muscles comprising quadriceps femoris.) Lower half of intertrochanteric line, linea aspera medial supracondylar line, medial intermuscular septum, tendon of adductor magnus

INSERTION: Quadriceps femoris tendon, medial border of patella, capsule of knee joint, by ligamentum patellae into tibial tuberosity

FUNCTION: Extends leg and draws patella medially

NERVE: Muscular branches of femoral

ARTERY: Muscular branches of femoral, muscular branches of profunda femoris, genicular branches of popliteal

References

	GRAY	GRANT'S ATLAS
Muscle	562-563	4-28A
Nerve	563, 1226, 1233	4-25
Artery	765, 769, 770	Not shown

VASTUS INTERMEDIUS*

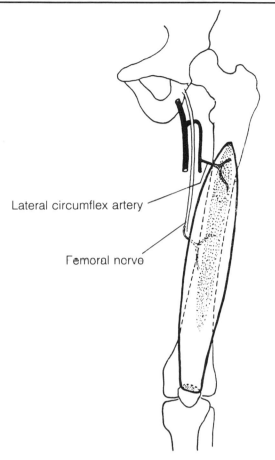

Lateral circumflex artery

Femoral nerve

ORIGIN: (Vastus Intermedius is 1 of 4 muscles comprising quadriceps femoris.) Upper two-thirds of anterior and lateral surface of femur, lower half of linea aspera, upper part of lateral supracondylar line, lateral intermuscular septum

INSERTION: Deep surface of tendons of rectus and vasti muscles, by ligamentum patellae into tibial tuberosity

FUNCTION: Extends leg

NERVE: Muscular branches of femoral

ARTERY: Lateral femoral circumflex

References

	GRAY	GRANT'S ATLAS
Muscle	**562-563**	**4-28B**
Nerve	**563, 1226, 1233**	**4-25**
Artery	**767**	**4-25**

*Articularis genus: a few separate muscle bundles arising deep to V. intermedius; tenses capsule of knee joint.

GRACILIS

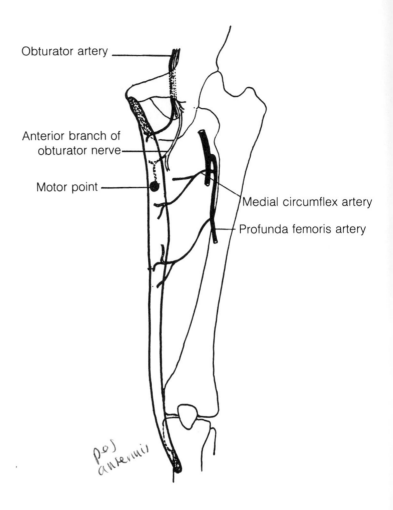

Obturator artery

Anterior branch of
obturator nerve

Motor point

Medial circumflex artery

Profunda femoris artery

pos
anterius

ORIGIN: Lower half of pubic symphysis, upper half of pubic arch
INSERTION: Upper part of medial surface of tibia
FUNCTION: Flexes leg and medially rotates at hip joint; adducts thigh
NERVE: Anterior branch of obturator
ARTERY: Muscular branches of profunda femoris, obturator, medial fem-
                   oral circumflex

References

	GRAY	GRANT'S ATLAS
Muscle	563	4-30
Nerve	563, 1226, 1230	4-25
Artery	753, 767	4-25

PECTINEUS

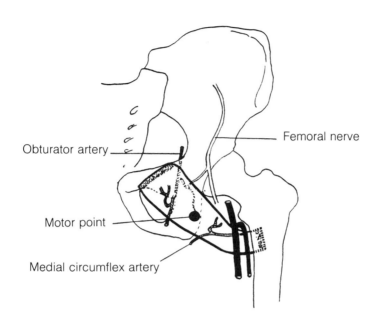

Femoral nerve

Obturator artery

Motor point

Medial circumflex artery

ORIGIN: Pectineal line, surface of pubis between iliopectineal eminence
               and pubic tubercle
INSERTION: Line extending from lesser trochanter to linea aspera
FUNCTION: Adducts, flexes, medially rotates thigh
NERVE: Muscular branches of femoral
ARTERY: Medial femoral circumflex, obturator

References

	GRAY	GRANT'S ATLAS
Muscle	563	4-22
Nerve	563, 1226, 1232	4-25
Artery	753, 767	4-25

ADDUCTOR LONGUS

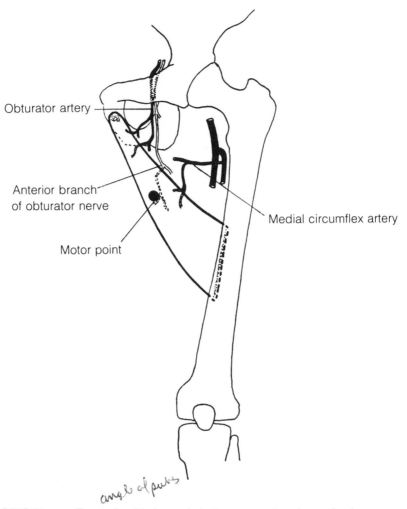

Obturator artery

Anterior branch
of obturator nerve

Motor point

Medial circumflex artery

angle of pubis

ORIGIN: Front of pubis in angle between crest and symphysis
INSERTION: Middle half of medial lip of linea aspera
FUNCTION: Adducts thigh, and assists in flexing it. Rotator action is contro-
                    versial
NERVE: Anterior branch of obturator
ARTERY: Medial femoral circumflex, obturator

References

	GRAY	GRANT'S ATLAS
Muscle	564	4-22
Nerve	564, 1226, 1230	4-25
Artery	753, 767	4-25

ADDUCTOR BREVIS

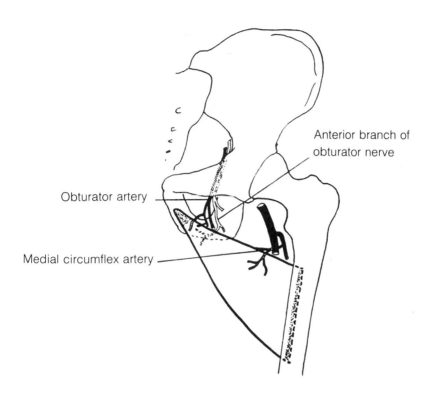

Anterior branch of
obturator nerve

Obturator artery

Medial circumflex artery

ORIGIN: Outer surface of inferior ramus of pubis *pectineal line*

INSERTION: Line extending from lesser trochanter to linea aspera

FUNCTION: Adducts thigh, and assists in flexing it. Rotator action is controversial

NERVE: Anterior branch of obturator

ARTERY: Medial femoral circumflex, obturator

References

	GRAY	GRANT'S ATLAS
Muscle	564	4-28B
Nerve	564, 1226, 1230	Not shown
Artery	753, 767	4-25

ADDUCTOR MAGNUS

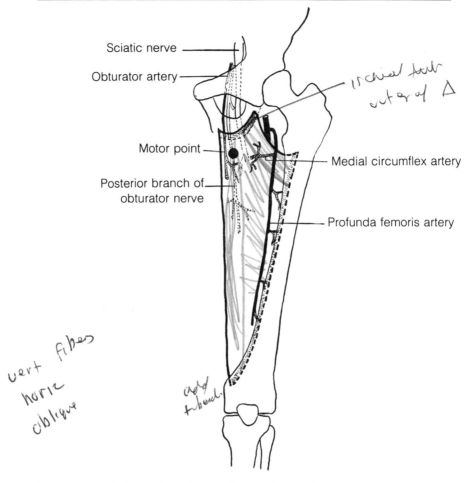

Sciatic nerve

Obturator artery

Motor point

Posterior branch of obturator nerve

ischial tub outer of △

Medial circumflex artery

Profunda femoris artery

vert fibers
horiz
oblique

add tuberd.

ORIGIN: Ischial tuberosity, rami of ischium and pubis

INSERTION: Line extending from greater trochanter to linea aspera, linea aspera, medial supracondylar line, adductor tubercle

FUNCTION: Adducts thigh; upper portion flexes it; lower portion extends it. Rotating action is controversial

NERVE: Posterior branch of obturator, sciatic

ARTERY: Medial femoral circumflex, perforating branches of profunda femoris, obturator, muscular branches of popliteal

References

	GRAY	GRANT'S ATLAS
Muscle	565	4-32A
Nerve	565, 1226, 1231, 1239	4-34
Artery	753, 767, 768	4-25, 4-34

GLUTEUS MAXIMUS

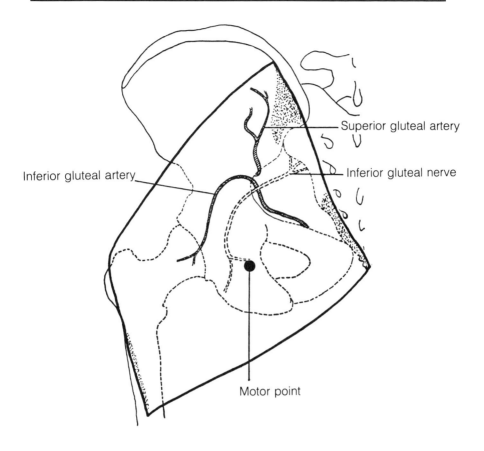

Superior gluteal artery

Inferior gluteal artery

Inferior gluteal nerve

Motor point

ORIGIN: Posterior gluteal line, tendon of sacrospinalis, dorsal surface of sacrum and coccyx, sacrotuberous ligament

INSERTION: Gluteal tuberosity of femur, iliotibial tract

FUNCTION: Extends thigh, assists in adducting and laterally rotating it; acting on insertion, muscle extends trunk

NERVE: Inferior gluteal

ARTERY: Superior gluteal, inferior gluteal, 1st perforating branch of profunda femoris

References

	GRAY	**GRANT'S ATLAS**
Muscle	566	4-31
Nerve	567, 1235, 1236	4-33, 4-34
Artery	757, 758, 768	4-33, 4-34

GLUTEUS MEDIUS

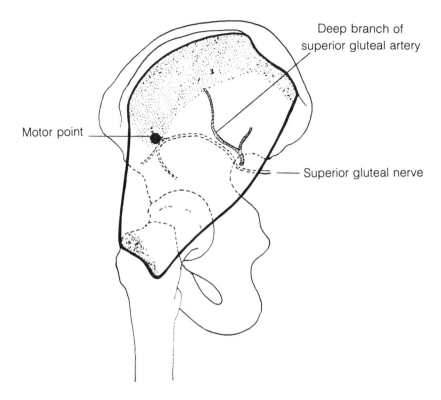

Deep branch of
superior gluteal artery

Motor point

Superior gluteal nerve

ORIGIN: Outer surface of ilium from iliac crest and posterior gluteal line above, to anterior gluteal line below, gluteal aponeurosis

INSERTION: Lateral surface of greater trochanter

FUNCTION: Abducts thigh, rotates thigh medially when limb is extended

NERVE: Superior gluteal

ARTERY: Deep branch of superior gluteal

References

	GRAY	GRANT'S ATLAS
Muscle	567	4-31
Nerve	568, 1235, 1236	4-34
Artery	757	4-33

GLUTEUS MINIMUS

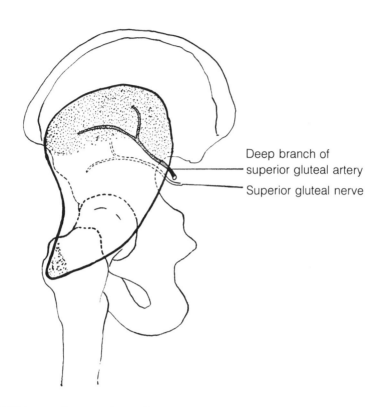

Deep branch of
superior gluteal artery

Superior gluteal nerve

ORIGIN: Outer surface of Illum between anterior and inferior gluteal lines
margin of greater sciatic notch

INSERTION: Anterior border of greater trochanter

FUNCTION: Abducts thigh, rotates thigh medially when limb is extended

NERVE: Superior gluteal

ARTERY: Deep branch of superior gluteal

References

	GRAY	**GRANT'S ATLAS**
Muscle	**568**	**4-34**
Nerve	**568, 1235, 1236**	**4-34**
Artery	**757**	**4-34**

TENSOR FASCIAE LATAE

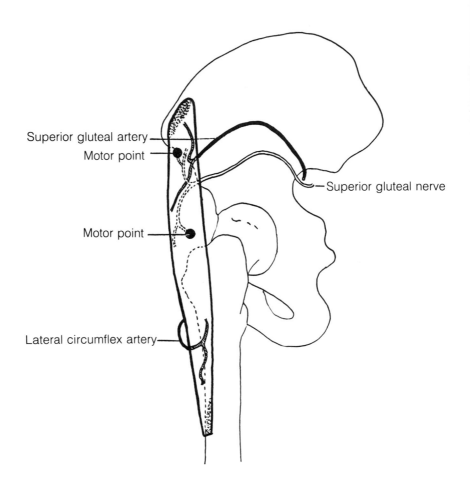

Superior gluteal artery
Motor point
Superior gluteal nerve
Motor point
Lateral circumflex artery

ORIGIN: Anterior part of outer lip of iliac crest, anterior border of ilium

INSERTION: Middle third of thigh along iliotibial tract

FUNCTION: Tenses fascia lata counteracting backward pull of gluteus maximus on iliotibial tract; assists in flexing, abducting, and medially rotating thigh

NERVE: Superior gluteal

ARTERY: Lateral femoral circumflex, superior gluteal

References

	GRAY	GRANT'S ATLAS
Muscle	568	4-22
Nerve	568, 1235, 1236	Not shown
Artery	757, 767	4-22

PIRIFORMIS

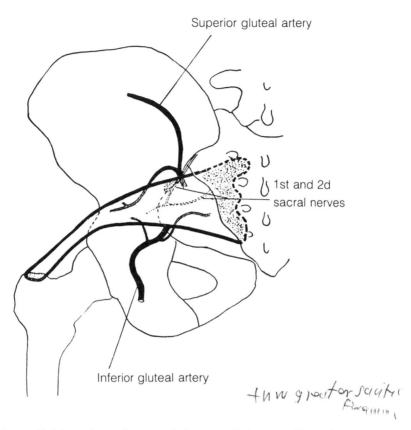

Superior gluteal artery

1st and 2d sacral nerves

Inferior gluteal artery

thru greater sciatic foramen

ORIGIN: Pelvic surface of sacrum between anterior sacral foramina, margin of greater sciatic foramen, sacrotuberous ligament
INSERTION: Upper border of greater trochanter of femur
FUNCTION: Rotates thigh laterally, abducts thigh when limb is flexed
NERVE: 1st and 2d sacral
ARTERY: Superior gluteal, inferior gluteal, internal pudendal

References

	GRAY	GRANT'S ATLAS
Muscle	568	4-36
Nerve	568, 1235, 1236	3-73
Artery	755, 757, 758	4-33, 4-34

OBTURATOR INTERNUS

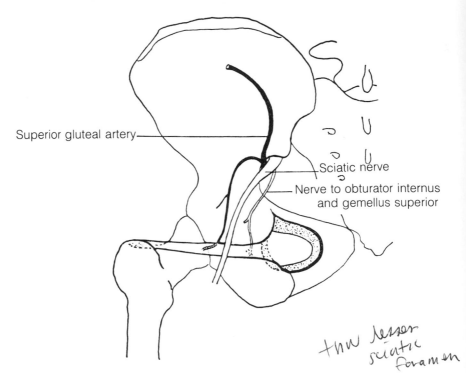

Superior gluteal artery—

Sciatic nerve

Nerve to obturator internus and gemellus superior

thru lesser sciatic foramen

ORIGIN: Margins of obturator foramen, obturator membrane, pelvic surface of hip bone behind and above obturator foramen, obturator fascia

INSERTION: Medial surface of greater trochanter

FUNCTION: Rotates thigh laterally, abducts thigh when limb is flexed

NERVE: Nerve to obturator internus and gemellus superior

ARTERY: Muscular branches of internal pudendal; superior gluteal

References

	GRAY	GRANT'S ATLAS
Muscle	568	3-55
Nerve	570, 1235	4-33
Artery	755, 757	3-55

GEMELLUS SUPERIOR

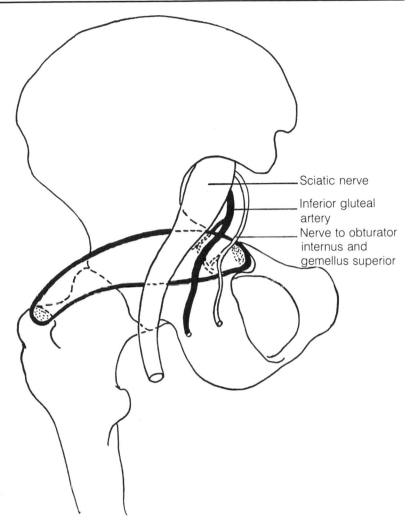

Sciatic nerve

Inferior gluteal artery

Nerve to obturator internus and gemellus superior

ORIGIN: Outer surface of ischial spine

INSERTION: Medial surface of greater trochanter, blends with obturator internus tendon

FUNCTION: Rotates thigh laterally

NERVE: Nerve to obturator internus and gemellus superior

ARTERY: Inferior gluteal

References

	GRAY	GRANT'S ATLAS
Muscle	570	4-37
Nerve	570, 1235	4-33, 4-34
Artery	758	4-33, 4-34

GEMELLUS INFERIOR

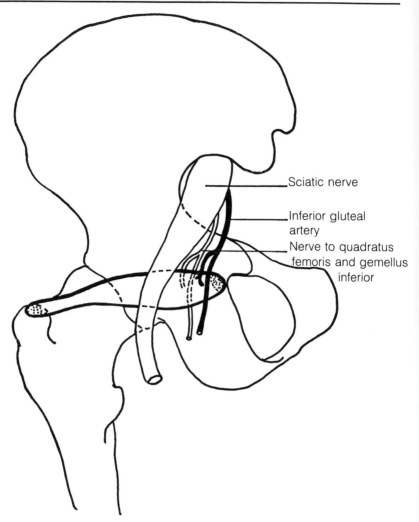

Sciatic nerve

Inferior gluteal artery

Nerve to quadratus femoris and gemellus inferior

ORIGIN: Upper part of ischial tuberosity
INSERTION: Medial surface of greater trochanter, blends with obturator internus tendon
FUNCTION: Rotates thigh laterally
NERVE: Nerve to quadratus femoris and gemellus inferior
ARTERY: Inferior gluteal

References

	GRAY	GRANT'S ATLAS
Muscle	570	4-37
Nerve	570, 1235	Not shown
Artery	758	4-33, 4-34

QUADRATUS FEMORIS

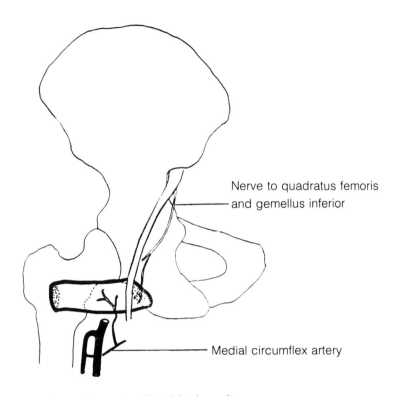

Nerve to quadratus femoris and gemellus inferior

Medial circumflex artery

ORIGIN: Lateral margin of ischial tuberosity
INSERTION: Quadrate tubercle of femur, linea quadrata
FUNCTION. Adducts and laterally rotates thigh
NERVE: Nerve to quadratus femoris and gemellus inferior
ARTERY: Medial femoral circumflex

References

	GRAY	GRANT'S ATLAS
Muscle	570	4-36
Nerve	570, 1235	Not shown
Artery	767	4-33, 4-34

OBTURATOR EXTERNUS

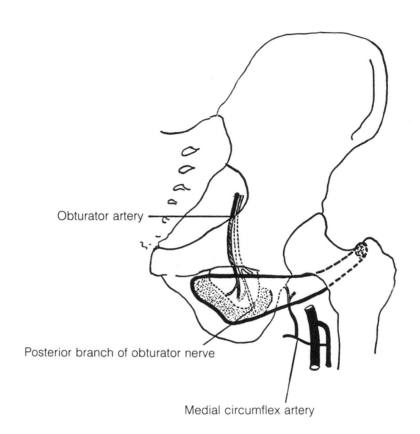

Obturator artery

Posterior branch of obturator nerve

Medial circumflex artery

ORIGIN: Outer margin of obturator foramen, outer surface of obturator membrane

INSERTION: Trochanteric fossa of femur

FUNCTION: Adducts thigh, rotates it laterally

NERVE: Posterior branch of obturator

ARTERY: Obturator, medial femoral circumflex

References

	GRAY	GRANT'S ATLAS
Muscle	570	4-42
Nerve	570, 1226, 1231	4-42
Artery	753, 767	4-34

BICEPS FEMORIS

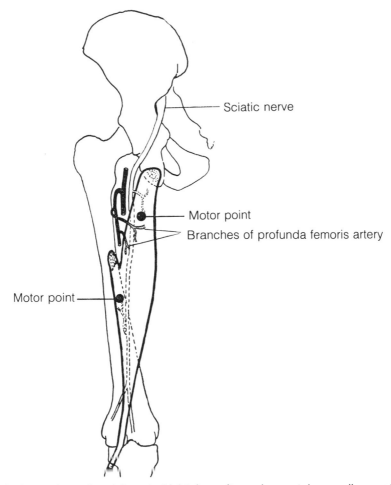

Sciatic nerve

Motor point

Branches of profunda femoris artery

Motor point

ORIGIN: Long head from ischial tuberosity and sacrotuberous ligament; short head from lateral lip of linea aspera, lateral supracondylar line of femur, lateral intermuscular septum

INSERTION: Head of fibula, lateral condyle of tibia, deep fascia on lateral side of leg

FUNCTION: Flexes leg, extends thigh, rotates leg laterally when knee is semi-flexed

NERVE: Sciatic (tibial portion to long head, peroneal portion to short head)

ARTERY: Perforating branches of profunda femoris, superior muscular branches of popliteal

References

	GRAY	GRANT'S ATLAS
Muscle	571	4-31, 4-32A
Nerve	572, 1235, 1239	4-33
Artery	768, 770	4-33, 4-34

SEMITENDINOSUS

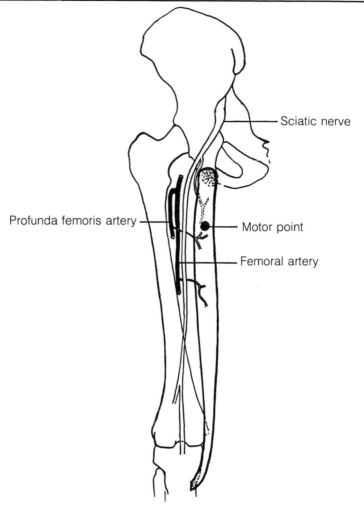

Sciatic nerve

Profunda femoris artery

Motor point

Femoral artery

ORIGIN: Upper and medial impression of ischial tuberosity with tendon of biceps

INSERTION: Upper part of medial surface of tibia, deep fascia of leg

FUNCTION: Flexes leg, extends thigh, rotates leg medially when knee is semiflexed

NERVE: Sciatic

ARTERY: Perforating branches of profunda femoris; superior muscular branches of popliteal   *inf gluteal*

References

	GRAY	GRANT'S ATLAS
Muscle	572	4-31
Nerve	572, 1235, 1239	4-33
Artery	768, 770	Not shown

SEMIMEMBRANOSUS

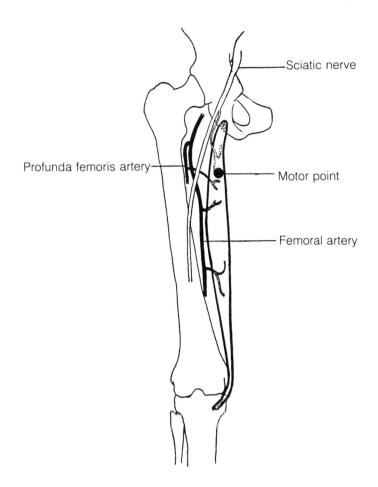

Sciatic nerve

Profunda femoris artery

Motor point

Femoral artery

ORIGIN: Upper and lateral facet of ischial tuberosity
INSERTION: Medial posterior surface of medial condyle of tibia
FUNCTION: Flexes leg, extends thigh, rotates leg medially when knee is
                    semiflexed
NERVE: Sciatic
ARTERY: Perforating branches of profunda femoris; superior muscular
                    branches of popliteal *inf gluteal*

References

	GRAY	GRANT'S ATLAS
Muscle	572	4-31
Nerve	572, 1235, 1239	4-33
Artery	768, 770	4-34

TIBIALIS ANTERIOR

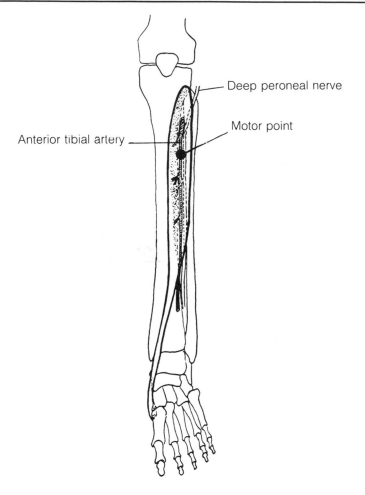

Deep peroneal nerve

Motor point

Anterior tibial artery

ORIGIN: Lateral condyle of tibia, upper two-thirds of lateral surface of tibia, interosseus membrane, deep fascia, lateral intermuscular septum

INSERTION: Medial and plantar surface of medial cuneiform bone, base of 1st metatarsal bone

FUNCTION: Dorsiflexes foot, inverts it

NERVE: Deep peroneal (anterior tibial)

ARTERY: Muscular branches of anterior tibial

References

	GRAY	GRANT'S ATLAS
Muscle	573	4-73, 4-98
Nerve	574, 1241	4-73
Artery	774	4-73

EXTENSOR HALLUCIS LONGUS

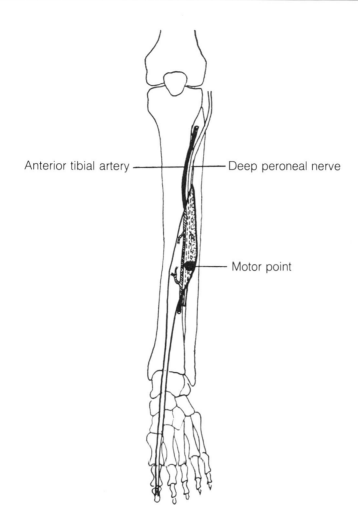

Anterior tibial artery — — Deep peroneal nerve

— Motor point

ORIGIN: Middle half of anterior surface of fibula, adjacent interosseous membrane
INSERTION: Base of distal phalanx of great toe
FUNCTION: Extends great toe, continued action dorsiflexes foot
NERVE: Deep peroneal (anterior tibial)
ARTERY: Muscular branches of anterior tibial

References

	GRAY	GRANT'S ATLAS
Muscle	574	4-73, 4-77
Nerve	575, 1241	4-73
Artery	774	4-73

EXTENSOR DIGITORUM LONGUS

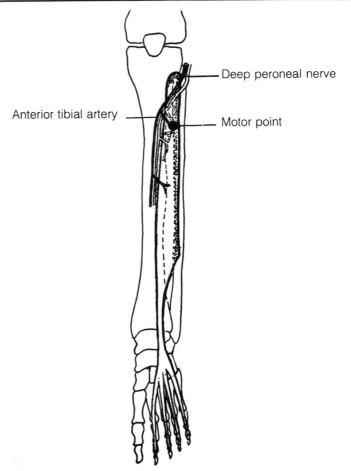

Deep peroneal nerve

Anterior tibial artery

Motor point

ORIGIN: Lateral condyle of tibia, upper three-fourths of anterior surface of fibula interosseous membrane, deep fascia, intermuscular septa

INSERTION: Dorsal surface of middle and distal phalanges of lateral 4 toes

FUNCTION: Extends phalanges of lateral 4 toes, continued action dorsiflexes foot

NERVE: Deep peroneal (anterior tibial)

ARTERY: Muscular branches of anterior tibial

References

	GRAY	GRANT'S ATLAS
Muscle	575	4-73, 4-77
Nerve	575, 1241	4-73
Artery	774	4-73

90

PERONEUS TERTIUS

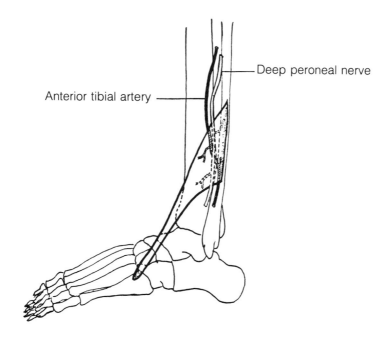

Deep peroneal nerve

Anterior tibial artery

ORIGIN: (Peroneus tertius is commonly known as the 5th tendon of extensor digitorum longus.) Lower anterior surface of fibula, adjacent intermuscular septum

INSERTION: Dorsal surface of base of 5th metatarsal bone

FUNCTION: Dorsiflexes and everts foot

NERVE: Deep peroneal (anterior tibial)

ARTERY: Muscular branches of anterior tibial

References

	GRAY	GRANT'S ATLAS
Muscle	575	4-71A, 4-73, 4-78D, 4-79
Nerve	575, 1241	4-73
Artery	774	Not shown

GASTROCNEMIUS

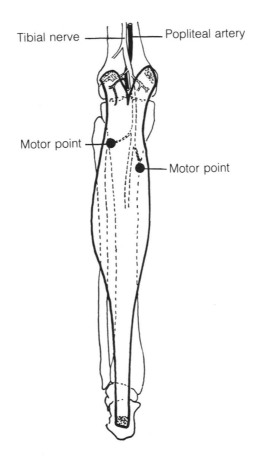

Tibial nerve —— —— Popliteal artery

Motor point —— ●

● —— Motor point

ORIGIN: Medial head from medial condyle and adjacent part of femur,
               capsule of knee joint; lateral head from lateral condyle and ad-
               jacent part of femur, capsule of knee joint
INSERTION: Into calcaneus by calcaneal tendon
FUNCTION: Plantarflexes foot; acting from below, it flexes femur on tibia
NERVE: Tibial (medial popliteal)
ARTERY: Sural branches of popliteal

References

	GRAY	GRANT'S ATLAS
Muscle	576	4-82A
Nerve	576, 1239	4-52, 4-82A
Artery	770	4-53

SOLEUS

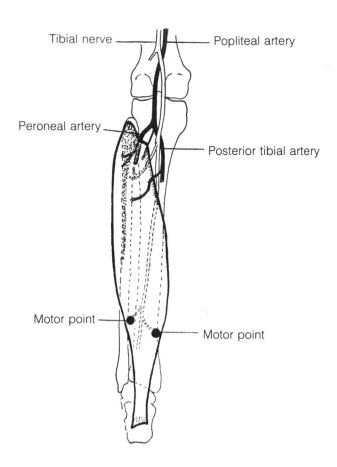

Tibial nerve — Popliteal artery

Peroneal artery — Posterior tibial artery

Motor point — Motor point

ORIGIN: Posterior surface of head and upper third of shaft of fibula, middle third of medial border of tibia, tendinous arch between tibia and fibula

INSERTION: Into calcaneus by calcaneal tendon

FUNCTION: Plantarflexes foot, steadies leg upon foot

NERVE: Tibial (medial popliteal)

ARTERY: Posterior tibial, peroneal, sural branches of popliteal

References

	GRAY	GRANT'S ATLAS
Muscle	576	4-82A, 4-83
Nerve	577, 1239	4-52, 4-83
Artery	770, 777, 779	4-83

PLANTARIS

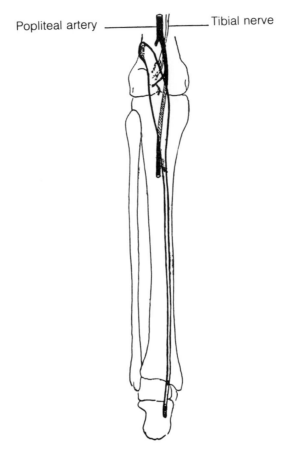

Popliteal artery ———————— Tibial nerve

ORIGIN: Lateral supracondylar line of femur, oblique popliteal ligament of knee joint

INSERTION: Medial side of posterior part of calcaneus, calcaneal tendon

FUNCTION: Plantarflexes foot

NERVE: Tibial (medial popliteal)

ARTERY: Sural branches of popliteal

References

	GRAY	GRANT'S ATLAS
Muscle	577	4-52
Nerve	577, 1239	Not shown
Artery	770	Not shown

POPLITEUS

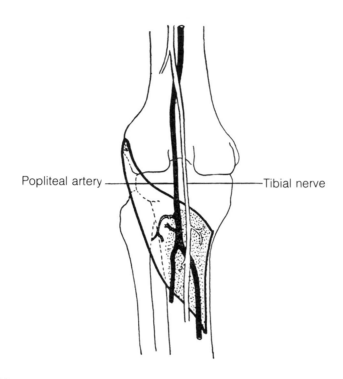

Popliteal artery ——————————————— Tibial nerve

ORIGIN: Lateral condyle of femur, oblique popliteal ligament of knee
INSERTION: Triangular area on posterior surface of tibia above soleal line
FUNCTION: Flexes leg, rotates tibia medially at beginning of flexion
NERVE: Tibial (medial or internal popliteal)
ARTERY: Genicular branches of popliteal

References

	GRAY	GRANT'S ATLAS
Muscle	577	4-53, 4-68, 4-86
Nerve	577, 1239	4-52, 4-86
Artery	772	4-53

FLEXOR HALLUCIS LONGUS

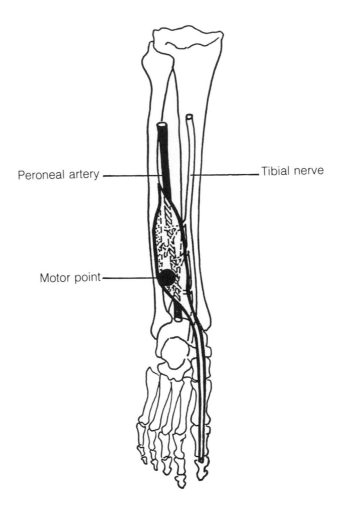

Peroneal artery —————— ———— Tibial nerve

Motor point—

ORIGIN: Lower two-thirds of posterior surface of fibula, interosseous membrane, adjacent intermuscular septa and fascia

INSERTION: Base of distal phalanx of great toe

FUNCTION: Flexes great toe, continued action aids in plantarflexing foot

NERVE: Tibial (medial or internal popliteal)

ARTERY: Muscular branches of peroneal

References

	GRAY	GRANT'S ATLAS
Muscle	578	4--84, 4-89
Nerve	578, 1239	4-86
Artery	777	4-84

FLEXOR DIGITORUM LONGUS

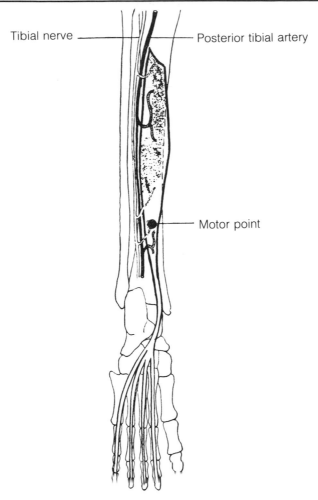

Tibial nerve ——————————— Posterior tibial artery

——— Motor point

ORIGIN: Posterior surface of middle three-fifths of tibia, fascia covering
            tibialis posterior
INSERTION: Plantar surface of base of distal phalanx of lateral 4 toes
FUNCTION: Flexes phalanges of lateral 4 toes, continued action plantarflexes
            foot
NERVE: Tibial (medial or internal popliteal)
ARTERY: Posterior tibial

References

	GRAY	GRANT'S ATLAS
Muscle	578	4-86, 4-95
Nerve	579, 1239	4-86
Artery	779	4-84

TIBIALIS POSTERIOR

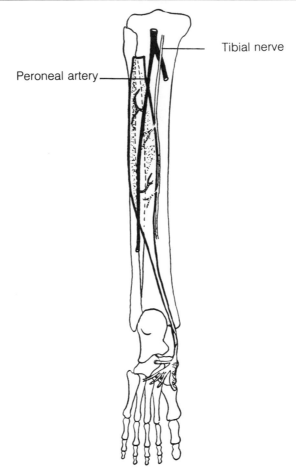

Peroneal artery

Tibial nerve

ORIGIN: Lateral part of posterior surface of tibia, upper two-thirds of medial surface of fibula, deep transverse fascia, adjacent intermuscular septa, posterior surface of interosseus membrane

INSERTION: Tuberosity of navicular bone, plantar surface of all cuneiform bones, plantar surface of base of 2d, 3d, and 4th metatarsal bones, cuboid bone, sustentaculum tali

FUNCTION: Plantarflexes foot, inverts it

NERVE: Tibial (medial or internal popliteal)

ARTERY: Peroneal

References

	GRAY	GRANT'S ATLAS
Muscle	579	4-86, 4-98
Nerve	579, 1239	4-86
Artery	777	4-86

PERONEUS LONGUS

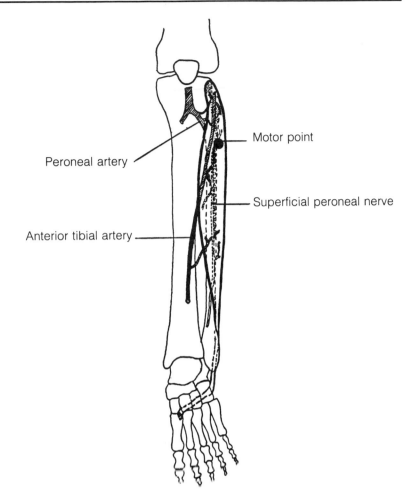

Motor point

Peroneal artery

Superficial peroneal nerve

Anterior tibial artery

ORIGIN: Lateral condyle of tibia, head and upper two-thirds of lateral surface of fibula, adjacent fascia, intermuscular septa

INSERTION: Lateral side of medial cuneiform bone, base of 1st metatarsal bone

FUNCTION: Plantarflexes foot, everts it

NERVE: Superficial peroneal (musculocutaneous)

ARTERY: Muscular branches of anterior tibial, muscular branches of peroneal

References

	GRAY	GRANT'S ATLAS
Muscle	579	**4-71A, 4-78D, 4-117**
Nerve	580, 1243	**4-75**
Artery	777	**Not shown**

PERONEUS BREVIS

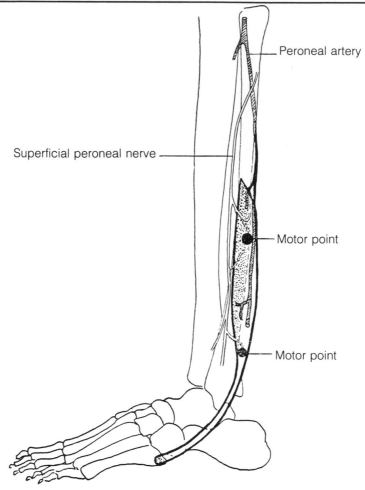

Peroneal artery

Superficial peroneal nerve

Motor point

Motor point

ORIGIN: Lower two-thirds of lateral surface of fibula, adjacent intermuscular septa

INSERTION: Lateral side of base of 5th metatarsal bone

FUNCTION: Plantarflexes foot, everts it

NERVE: Superficial peroneal (musculocutaneous)

ARTERY: Muscular branches of peroneal

References

	GRAY	GRANT'S ATLAS
Muscle	580	4-71A, 4-78D
Nerve	580, 1243	4-75
Artery	777	Not shown

EXTENSOR DIGITORUM BREVIS

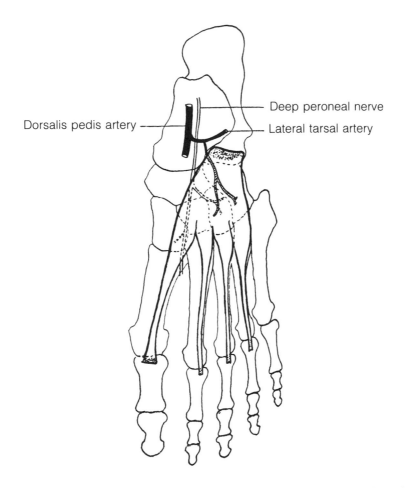

Deep peroneal nerve

Dorsalis pedis artery

Lateral tarsal artery

ORIGIN: Forepart of upper and lateral surface of calcaneus, lateral talo-calcaneal ligament, cruciate crural ligament

INSERTION: 1st tendon into dorsal surface of base of proximal phalanx of great toe, remaining 3 tendons into lateral sides of tendons of extensor digitorum longus

FUNCTION: Extends phalanges of 4 medial toes

NERVE: Deep peroneal (anterior tibial)

ARTERY: Dorsalis pedis, lateral tarsal

References

	GRAY	GRANT'S ATLAS
Muscle	584	4-77, 4-79
Nerve	584, 1241	4-73
Artery	775	Not shown

ABDUCTOR HALLUCIS

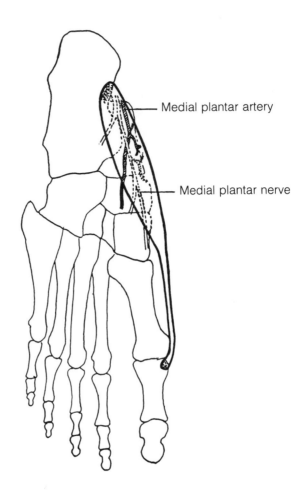

Medial plantar artery

Medial plantar nerve

ORIGIN: Medial process of calcaneus, laciniate ligament, plantar aponeurosis, adjacent intermuscular septum; tub of navicular

INSERTION: Medial side of base of proximal phalanx of great toe

FUNCTION: Abducts great toe

NERVE: Medial plantar

ARTERY: Medial plantar

References

	GRAY	GRANT'S ATLAS
Muscle	586	4-102
Nerve	586, 1240	4-87, 4-100
Artery	779	4-87

FLEXOR DIGITORUM BREVIS

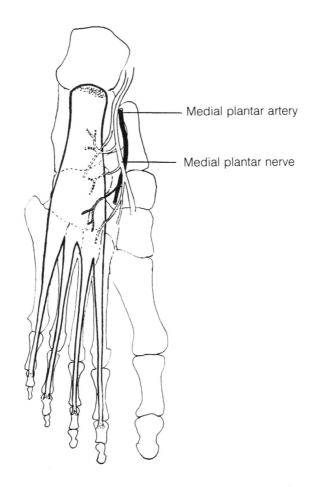

— Medial plantar artery

— Medial plantar nerve

ORIGIN: Medial process of calcaneus, plantar aponeurosis, adjacent intermuscular septa

INSERTION: Middle phalanx of lateral 4 toes

FUNCTION: Flexes middle phalanges on proximal, continued action also flexes proximal phalanges of lateral 4 toes

NERVE: Medial plantar

ARTERY: ~~Medial~~ plantar
Lateral

References

	GRAY	GRANT'S ATLAS
Muscle	**586**	**4-93**
Nerve	**586, 1240**	**Not shown**
Artery	**779**	**Not shown**

ABDUCTOR DIGITI MINIMI

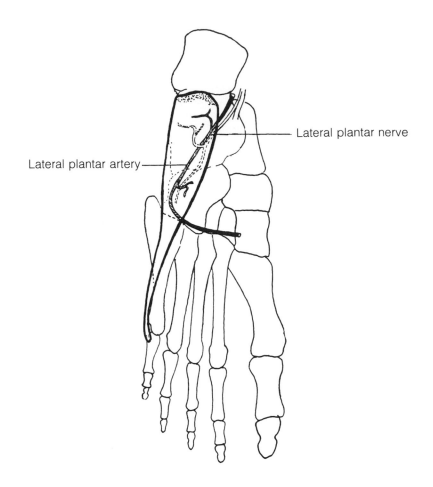

Lateral plantar nerve

Lateral plantar artery

ORIGIN: Lateral and medial processes of calcaneus, calcaneal fascia, adjacent intermuscular septum   *tub at base of 5th MT*

INSERTION: Lateral side of base of proximal phalanx of little toe

FUNCTION: Abducts little toe, assists in flexing it

NERVE: Lateral plantar

ARTERY: Lateral plantar

References

	GRAY	GRANT'S ATLAS
Muscle	586	4-102
Nerve	587, 1241	Not shown
Artery	779	Not shown

QUADRATUS PLANTAE
(Flexor Accessorius)

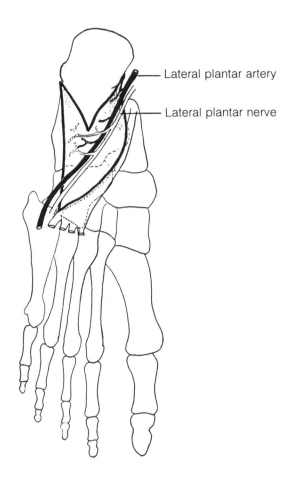

Lateral plantar artery

Lateral plantar nerve

ORIGIN: <u>Medial head</u> from medial surface of calcaneus and medial border of long plantar ligament, <u>lateral head</u> from lateral border of plantar surface of calcaneus and lateral border of long plantar ligament

INSERTION: Tendons of flexor digitorum longus

FUNCTION: Flexes terminal phalanges of lateral 4 toes

NERVE: Lateral plantar

ARTERY: Lateral plantar

References

	GRAY	GRANT'S ATLAS
Muscle	**587**	**4-95, 4-98**
Nerve	**587, 1241**	**4-100**
Artery	**779**	**4-100**

LUMBRICALES

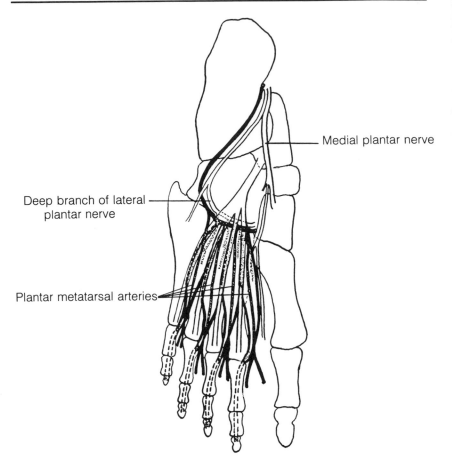

Medial plantar nerve

Deep branch of lateral plantar nerve

Plantar metatarsal arteries

ORIGIN: There are 4 lumbricales, all arising from tendons of flexor digitorum longus: 1st from medial side of tendon for 2d toe, 2d from adjacent sides of tendons for 2d and 3d toes, 3d from adjacent sides of tendons for 3d and 4th toes, 4th from adjacent sides of tendons for 4th and 5th toes

INSERTION: With tendons of extensor digitorum longus and interossei into bases of terminal phalanges of 4 lateral toes

FUNCTION: Flex toes at metatarsophalangeal joints, extend toes at interphalangeal joints

NERVE: Medial plantar, deep lateral plantar

ARTERY: Plantar metatarsal

References

	GRAY	GRANT'S ATLAS
Muscle	588	4-95
Nerve	588, 1240, 1241	Not shown
Artery	780	4-93

FLEXOR HALLUCIS BREVIS

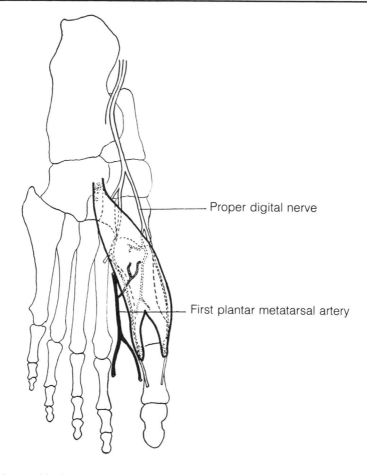

Proper digital nerve

First plantar metatarsal artery

ORIGIN: Medial part of plantar surface of cuboid bone, adjacent portion of lateral cuneiform bone, prolongation of tendon of tibialis posterior

INSERTION: Medial and lateral side of proximal phalanx of great toe

FUNCTION: Flexes great toe

NERVE: Proper digital nerve of great toe (1st plantar digital nerve) of medial plantar nerve

ARTERY: First plantar metatarsal (from junction of lateral and deep plantar arteries)

References

	GRAY	GRANT'S ATLAS
Muscle	588	4-102
Nerve	588, 1240	4-100
Artery	780	Not shown

ADDUCTOR HALLUCIS

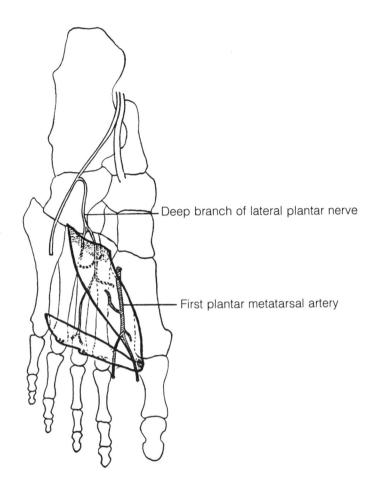

Deep branch of lateral plantar nerve

First plantar metatarsal artery

ORIGIN: Oblique head from bases of 2d, 3d, and 4th metatarsal bones, sheath of peroneus longus; transverse head from capsules of 2d, 3d, 4th, and 5th metatarsophalangeal ligaments, transverse ligament of sole

INSERTION: Lateral side of base of proximal phalanx of great toe

FUNCTION: Adducts great toe, assists in flexing it

NERVE: Deep branch of lateral plantar

ARTERY: First plantar metatarsal

References

	GRAY	GRANT'S ATLAS
Muscle	589	4-100
Nerve	589, 1241	4-100
Artery	780	Not shown

FLEXOR DIGITI MINIMI BREVIS

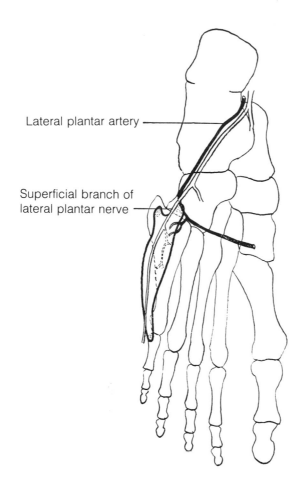

Lateral plantar artery ───

Superficial branch of
lateral plantar nerve ───

ORIGIN: Sheath of peroneus longus, base of 5th metatarsal bone
INSERTION: Lateral side of base of proximal phalanx of little toe
FUNCTION: Flexes little toe
NERVE: Superficial branch of lateral plantar
ARTERY: Lateral plantar

References

	GRAY	**GRANT'S ATLAS**
Muscle	589	4-102
Nerve	589, 1241	4-100
Artery	779	Not shown

INTEROSSEI DORSALES
(Dorsal Interossei)

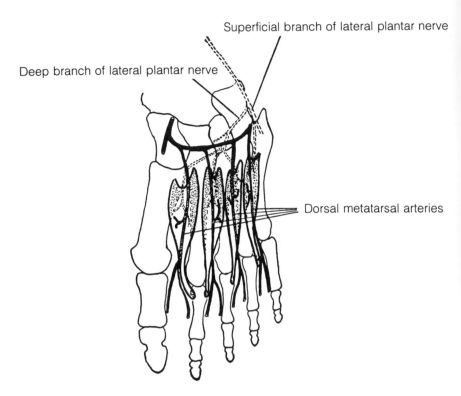

Superficial branch of lateral plantar nerve

Deep branch of lateral plantar nerve

Dorsal metatarsal arteries

ORIGIN: There are 4 dorsal interossei, each arising by 2 heads from adjacent sides of metatarsal bones

INSERTION: 1st into medial side of proximal phalanx of 2d toe, 2d into lateral side of proximal phalanx of 2d toe, 3d into lateral side of proximal phalanx of 3d toe, 4th into lateral side of proximal phalanx of 4th toe

FUNCTION: Abduct 2d, 3d, and 4th toes from axis of 2d toe, assist in flexing proximal phalanges and in extending middle and distal phalanges

NERVE: Superficial and deep branches of lateral plantar

ARTERY: Dorsal metatarsal

References

	GRAY	GRANT'S ATLAS
Muscle	589	4-77, 4-102
Nerve	590, 1241	4-100
Artery	776	4-75

INTEROSSEI PLANTARES
(Plantar Interossei)

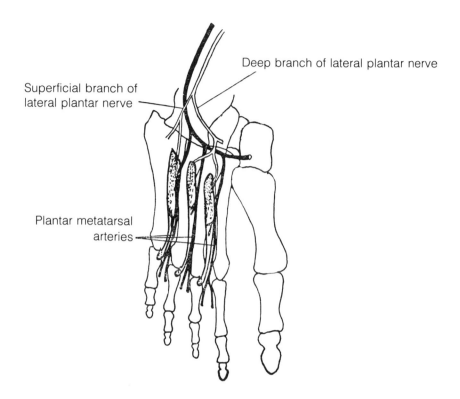

Deep branch of lateral plantar nerve

Superficial branch of
lateral plantar nerve

Plantar metatarsal
arteries

ORIGIN: There are 3 plantar interossoi, arising from bases and medial
sides of 3d, 4th, and 5th metatarsal bones

INSERTION: Medial sides of bases of proximal phalanges of 3d, 4th, and 5th
toes

FUNCTION: Adduct 3d, 4th, and 5th toes toward axis of 2d toe, assist in
flexing proximal phalanges and in extending middle and distal
phalanges

NERVE: Superficial and deep branches of lateral plantar

ARTERY: Plantar metatarsal

References

	GRAY	GRANT'S ATLAS
Muscle	590	4-102
Nerve	590, 1241	4-100
Artery	780	Not shown

SCHEME OF THE BRACHIAL PLEXUS

Nerves	Trunks	Divisions	Cords	Main Branches

(Redrawn after Cunningham)

NERVES OF UPPER EXTREMITIES

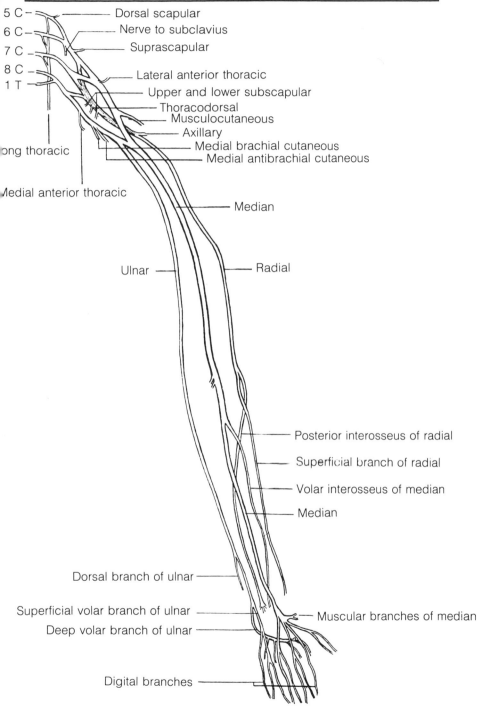

5 C — Dorsal scapular
6 C — Nerve to subclavius
7 C — Suprascapular
8 C — Lateral anterior thoracic
1 T — Upper and lower subscapular
Thoracodorsal
Musculocutaneous
Axillary
Medial brachial cutaneous
ong thoracic — Medial antibrachial cutaneous

Medial anterior thoracic

Median

Ulnar — Radial

Posterior interosseus of radial

Superficial branch of radial

Volar interosseus of median

Median

Dorsal branch of ulnar

Superficial volar branch of ulnar — Muscular branches of median
Deep volar branch of ulnar

Digital branches

113

ARTERIES OF UPPER EXTREMITIES

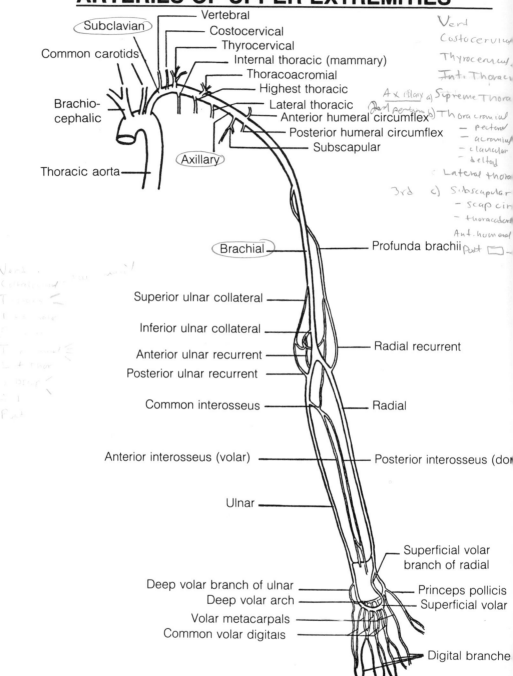

Subclavian

Common carotids

Vertebral
Costocervical
Thyrocervical
Internal thoracic (mammary)
Thoracoacromial
Highest thoracic

Brachio-
cephalic

Lateral thoracic
Anterior humeral circumflex
Posterior humeral circumflex
Subscapular

Thoracic aorta

Axillary

Brachial

Profunda brachii

Superior ulnar collateral

Inferior ulnar collateral

Anterior ulnar recurrent

Posterior ulnar recurrent

Radial recurrent

Common interosseus

Radial

Anterior interosseus (volar)

Posterior interosseus (do

Ulnar

Superficial volar
branch of radial

Deep volar branch of ulnar

Deep volar arch

Volar metacarpals

Common volar digitals

Princeps pollicis
Superficial volar

Digital branche

Vert
Costocervic
Thyrocervic
Inti. Thoraci
A x illary a) Supreme Thora
(2nd portion) b) Thoracromial
— pectoral
— acromial
— clavicular
— deltoid
Lateral thora
3rd c) Subscapular
— scap cir
— thoracodor
Ant. humeral
Post

NERVES OF LOWER EXTREMITIES

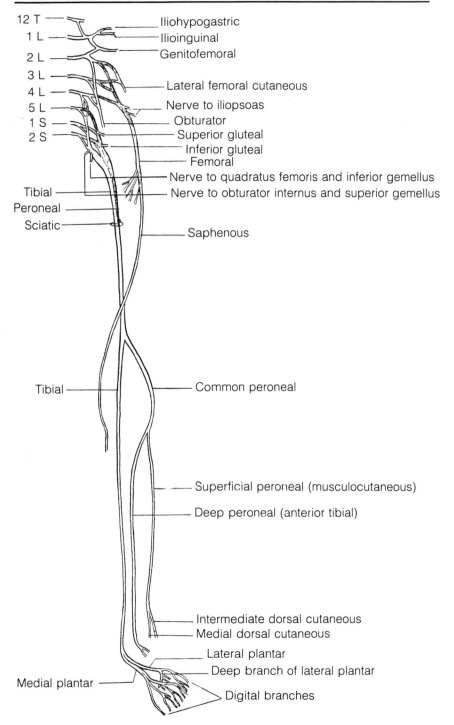

12 T — Iliohypogastric
1 L — Ilioinguinal
2 L — Genitofemoral
3 L —
4 L — Lateral femoral cutaneous
5 L — Nerve to iliopsoas
1 S — Obturator
2 S — Superior gluteal
Inferior gluteal
Femoral
Nerve to quadratus femoris and inferior gemellus
Tibial — Nerve to obturator internus and superior gemellus
Peroneal —
Sciatic — Saphenous

Tibial — Common peroneal

Superficial peroneal (musculocutaneous)

Deep peroneal (anterior tibial)

Intermediate dorsal cutaneous
Medial dorsal cutaneous
Lateral plantar
Deep branch of lateral plantar
Medial plantar — Digital branches

ARTERIES OF LOWER EXTREMITIES

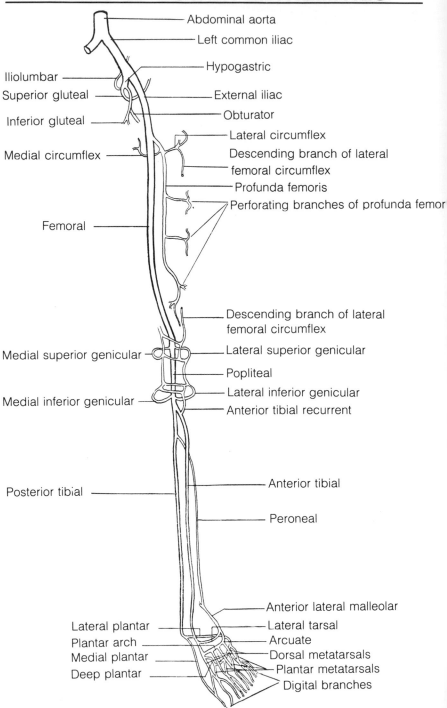

Abdominal aorta

Left common iliac

Hypogastric

Iliolumbar

Superior gluteal

External iliac

Inferior gluteal

Obturator

Lateral circumflex

Descending branch of lateral femoral circumflex

Medial circumflex

Profunda femoris

Perforating branches of profunda femor

Femoral

Descending branch of lateral femoral circumflex

Medial superior genicular

Lateral superior genicular

Popliteal

Lateral inferior genicular

Medial inferior genicular

Anterior tibial recurrent

Posterior tibial

Anterior tibial

Peroneal

Anterior lateral malleolar

Lateral plantar

Lateral tarsal

Plantar arch

Arcuate

Medial plantar

Dorsal metatarsals

Deep plantar

Plantar metatarsals

Digital branches

MUSCLES OF RIGHT UPPER EXTREMITY— ANTERIOR VIEW

ORIGINS—SOLID   INSERTIONS—STIPPLED

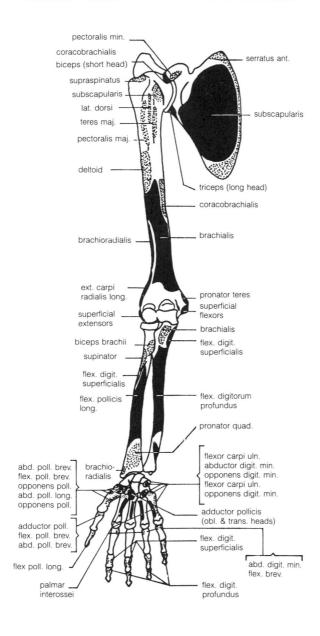

pectoralis min.
coracobrachialis
biceps (short head)
supraspinatus
subscapularis
lat. dorsi
teres maj.
pectoralis maj.
deltoid

serratus ant.
subscapularis
triceps (long head)
coracobrachialis
brachialis

brachioradialis

ext. carpi radialis long.
superficial extensors
biceps brachii
supinator
flex. digit. superficialis
flex. pollicis long.

pronator teres
superficial flexors
brachialis
flex. digit. superficialis
flex. digitorum profundus
pronator quad.

abd. poll. brev.
flex. poll. brev.
opponens poll.
abd. poll. long.
opponens poll.

brachio-radialis

flexor carpi uln.
abductor digit. min.
opponens digit. min.
flexor carpi uln.
opponens digit. min.

adductor pollicis (obl. & trans. heads)

adductor poll.
flex. poll. brev.
abd. poll. brev.

flex. digit. superficialis

flex poll. long.

abd. digit. min.
flex. brev.

palmar interossei

flex. digit. profundus

117

MUSCLES OF RIGHT UPPER EXTREMITY— POSTERIOR VIEW

ORIGINS—SOLID   INSERTIONS—STIPPLED

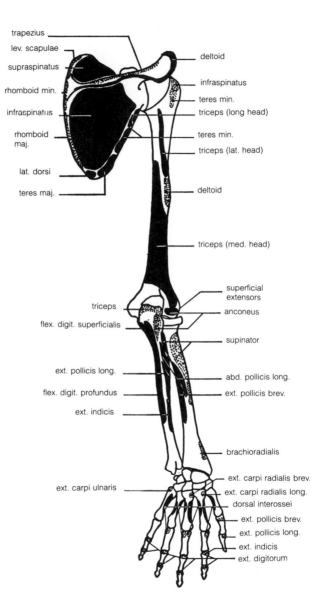

trapezius

lev. scapulae

supraspinatus

rhomboid min.

infraspinatus

rhomboid maj.

lat. dorsi

teres maj.

triceps

flex. digit. superficialis

ext. pollicis long.

flex. digit. profundus

ext. indicis

ext. carpi ulnaris

deltoid

infraspinatus

teres min.

triceps (long head)

teres min.

triceps (lat. head)

deltoid

triceps (med. head)

superficial extensors

anconeus

supinator

abd. pollicis long.

ext. pollicis brev.

brachioradialis

ext. carpi radialis brev.

ext. carpi radialis long.

dorsal interossei

ext. pollicis brev.

ext. pollicis long.

ext. indicis

ext. digitorum

MUSCLES OF RIGHT LOWER EXTREMITY—ANTERIOR VIEW

ORIGINS—SOLID   INSERTIONS—STIPPLED

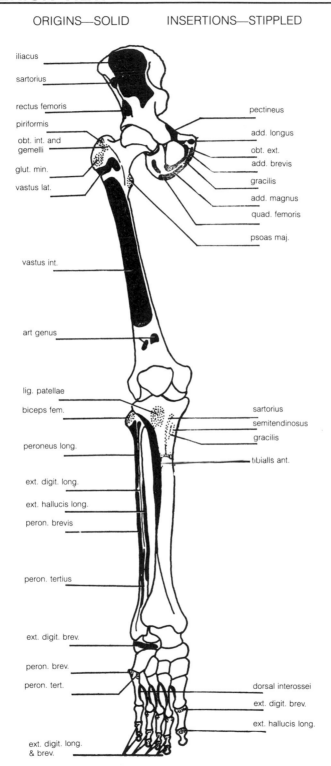

iliacus

sartorius

rectus femoris

piriformis

obt. int. and
gemelli

glut. min.

vastus lat.

vastus int.

art genus

lig. patellae

biceps fem.

peroneus long.

ext. digit. long.

ext. hallucis long.

peron. brevis

peron. tertius

ext. digit. brev.

peron. brev.

peron. tert.

ext. digit. long.
& brev.

pectineus

add. longus

obt. ext.

add. brevis

gracilis

add. magnus

quad. femoris

psoas maj.

sartorius

semitendinosus

gracilis

tibialis ant.

dorsal interossei

ext. digit. brev.

ext. hallucis long.

MUSCLES OF RIGHT LOWER EXTREMITY— POSTERIOR VIEW

ORIGINS—SOLID   INSERTIONS—STIPPLED

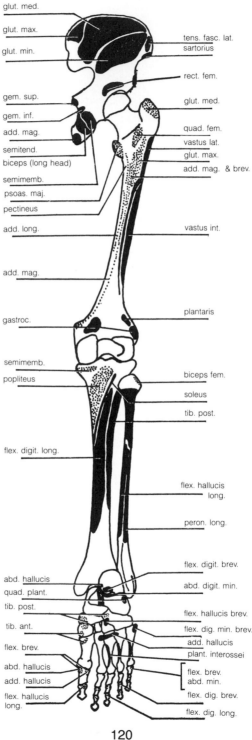

glut. med.

glut. max.

glut. min.

tens. fasc. lat.
sartorius

rect. fem.

gem. sup.

glut. med.

gem. inf.

add. mag.

quad. fem.

vastus lat.

semitend.

glut. max.

biceps (long head)

add. mag. & brev.

semimemb.

psoas. maj.

pectineus

add. long.

vastus int.

add. mag.

plantaris

gastroc.

semimemb.

biceps fem.

popliteus

soleus

tib. post.

flex. digit. long.

flex. hallucis long.

peron. long.

flex. digit. brev.

abd. hallucis

abd. digit. min.

quad. plant.

tib. post.

flex. hallucis brev.

tib. ant.

flex. dig. min. brev.

add. hallucis

flex. brev.

plant. interossei

abd. hallucis

flex. brev.
abd. min.

add. hallucis

flex. dig. brev.

flex. hallucis long.

flex. dig. long.

INDEX

CHARTS

MUSCLES